MW00523035

# COMPUTATIONAL DRUG DESIGN

# COMPUTATIONAL DRUG DESIGN

## A Guide for Computational and Medicinal Chemists

**DAVID C. YOUNG**
Computer Sciences Corporation

A JOHN WILEY & SONS, INC., PUBLICATION

Published by John Wiley & Sons, Inc., Hoboken, New Jersey
Published simultaneously in Canada



***Library of Congress Cataloging-in-Publication Data:***

Young, David C., 1964–
Computational drug design / David C. Young.
p.; cm.
Includes bibliographical references and index.
ISBN 978-0-470-12685-1 (cloth/CD)

1. Drugs—Design—Mathematical models. 2. Drugs—Design—Data processing.
I. Title.
[DNLM: 1. Computational Biology—methods. 2. Drug Design. 3. Biochemical Phenomena.
4. Chemistry, Pharmaceutical—methods. 5. Drug Delivery Systems.
6. Models, Chemical. QV 744 Y69c 2009]
RS420. Y68 2009
615′.190285--dc22 2008041828

Printed in the United States of America

10 9 8 7 6 5 4 3 2 1

*This book is dedicated to my grandfathers, Harvey Turner and Ray Young.*

*Harvey Turner had the intelligence to work his way up from a draftsman to Chief Engineer at Donaldsons. Then he had the wisdom to leave that high pressure career behind and spend the next two decades teaching art.*

*Ray Young dropped out of high school to help make ends meet during the great depression. He never returned to school, but was the most widely read and knowledgeable person I have every met.*

# CONTENTS

# PREFACE

*A pharmaceutical company utilizing computational drug design is like an organic chemist utilizing an NMR. It won't solve all of your problems, but you are much better off with it than without it.*

The design of a new drug is an incredibly difficult and frustrating task. If it weren't for the potential to earn equally incredible profits, the massive costs and aggravation over failed experiments would dissuade any reasonable person from undertaking such a career. There is no one scientific technique used to design a new pharmaceutical product. It is instead a collaborative process in which every available technique, and a few more invented on the spur of the moment, are utilized in order to achieve the desired results.

There are books that talk about drug design tools, algorithms, and mathematical functions, and books that give some results showing that one compound worked better than another for inhibiting a particular enzyme. However, these books spend surprisingly little time discussing the process that the chemist goes through to actually design a new drug molecule. This book is oriented around the way that computational techniques are utilized in the drug design process.

Typical design processes for a number of drug development scenarios are presented in the first part of the book. Multiple drug design processes are presented, because the process itself changes depending upon whether the drug target is a protein, DNA, a target within the central nervous system, etc. The design processes presented in this text do not reflect the process at any one specific pharmaceutical company, but are rather typical work flows incorporating the elements that are used in one way or another at almost all pharmaceutical research campuses. The chapters on the drug design process are intended to show how each of the computational techniques are typically utilized. The comparison of different drug design processes illustrates where specific computational tools would and would not be appropriate. The text presents many rules of thumb for choosing which tools are best utilized under certain situations.

The second part of the book has a series of chapters, each focusing on one computational technique. The chapters on each of the computational techniques are intended to give a solid understanding of the strengths and weaknesses of the method. The underlying theory is discussed in concept, but with little if any mathematical derivation. The processes for using the software and important issues that tend to arise are described. Where there are significant differences between available software packages, those issues are discussed. However, the text is not specific to one manufacturer's software. The relative merits of various methods are discussed, and, where possible, a table with quantitative comparisons is presented.

The third part of the book gives a few chapters discussing related topics. These are topics that drug design chemists should have some familiarity with, but are not usually engaged in on a daily basis. Fields of research so new that they are still being defined at the time this book was written are also introduced here. Since any detailed information on such subjects would be obsolete before the ink on this book is dry, some of these introductions are kept intentionally broad and conceptual.

In a book that covers a broad subject area, it is always difficult to choose which references to include for each chapter. For this text, I have taken a two-fold approach. Key references are listed at the end of each chapter in an annotated bibliography. These references tend to be the next place that readers should look for additional information on the topics discussed in the chapter. This is supplemented by a longer reference list included on the accompanying CD. Readers wishing to delve very deeply into a particular subject will find this larger list of references valuable.

This book is very industry-centric. The discussions of when and how tools are used is based on a typical pharmaceutical industry drug design process. As such, I have intentionally avoided using cartoon-like illustrations of geometric figures fitting together. The majority of the figures in this book are screen shots of actual software packages that drug designers might use on a daily basis. This is the environment that a drug designer in the pharmaceutical industry must learn to work in.

For students interested in pursuing a career in the drug design field, this text is intended to give an ideal starting point for their studies. The text assumes a solid background in chemistry, a basic understanding of biochemistry, and only minimal previous exposure to computational chemistry.

Researchers already employed in the drug design field will be particularly interested in the tables comparing accuracies of docking methods. There is also a fairly large table of bioisosteric substitutions. Providing an overview of the whole field may turn out to be this book's greatest contribution.

I wish you the best of success in pursuing your drug design activities.

David C. Young

# ACKNOWLEDGMENTS

There is a popular myth that books are written by solitary people typing away in a lonely, deserted house. Indeed, there are many hours spent in front of a keyboard. However, a book would never come into being without the help, support, and hard work of the author's family, colleagues, co-workers, editors, graphic artists, and random other people saying, "Wow, that sounds complicated."

My family has been exceptionally tolerant of my ever-present laptop in the car, during swimming lessons, in front of the TV, and at this very minute sitting off to the side as my wife Natalie displays her stained glass work at an art show. My oldest son Gregory is a man of very few words, but the occasional "cool dad" speaks volumes. My daughter Ariel thinks it is neat that her dad is a scientist, but still won't ask for help with her college freshman chemistry homework. My youngest, Isaac, has little interest in anything that doesn't involve video games or reptiles, but he seems to consider docking calculations with solvation and entropy corrections to be dad's form of video game.

My current job at the Alabama Supercomputer Center allows me the chance to interact with faculty and students of many different disciplines throughout the state of Alabama. Randy Fulmer and the Alabama Supercomputer Authority staff are always interested to hear about the scientific research utilizing the supercomputers here. I've had bosses both good and bad, and David Ivey at CSC is definitely the best. Charles Wright and Derek Gottlieb always think of me as the software guy. They won't forget the day they asked about quantum chemistry software and got way more than they bargained for (the rest of the staff is afraid to ask).

This is my second book with John Wiley & Sons. I wouldn't consider working with any other publisher as long as Wiley will have me. Anita Lekhwani and Rebecka Ramos have been wonderful to work with. There are many others at Wiley who contribute to creating a high quality book as they format the tables, integrate the artwork, and lovingly cover the manuscript in red ink.

Within the pharmaceutical field, I have had the pleasure of working with some excellent people. Andy Peek, now at Integrated DNA Technologies Inc., manages to be top notch at bioinformatics without succumbing to the high pressure of the drug design world. It has also been my privilege to work with Brad Poland, now at Pfizer, who is a wonderful crystallographer and co-worker. Working with Stephan Reiling, now at Aventis, and the entire SARNavigator development team was the most enjoyment I ever got from my job. Mitch Polley, who left Tripos to return home to Australia, has become a good friend as well as teaching me much about drug design.

I wanted the majority of the figures in the book to show commercial drug design software, which is the environment that drug designers must learn to work in. I greatly appreciate getting demo copies of software from ACD, Accelrys, Cambridge Crystallographic Data Centre, Chemical Computing Group, Conflex, COSMOlogic, SimBioSys, Simulations Plus, Tripos, and Wavefunction for this purpose. Those same companies were invited to contribute product literature, white papers, and demo software to the accompanying CD.

David C. Young

# ABOUT THE AUTHOR

David Young's career has taken him to the far corners of computational chemistry. He was assistant director of drug design for a now nonexistent startup company, eXegenics. David once taught introductory science and graduate computer programming courses at Auburn University. He has also written quite a bit of software for Tripos and others. Dr Young is currently employed by Computer Sciences Corporation (CSC) as a chemistry software expert, under contract to the Alabama Supercomputer Authority. Much earlier in his life, he ran the nuclear reactor aboard a ballistic missile submarine.

Dr Young received his PhD in chemistry from Michigan State University, under the direction of James Harrison. He also has degrees in computational mathematics and business. His previous book, *Computational Chemistry: A Practical Guide for Applying Techniques to Real World Problems*, has been on the John Wiley & Sons bestseller list.

David currently lives in Huntsville, Alabama, where he provides software technical support to users of the Alabama Supercomputer Center.

# SYMBOLS AND ACRONYMS USED IN THIS BOOK

| | |
|---|---|
| 1D | one-dimensional |
| 2D | two-dimensional |
| 2.5D | two-and-a-half-dimensional |
| 3D | three-dimensional |
| 3D-QSAR | three-dimensional quantitative structure–activity relationship |
| Å | ångström |
| ACD | Advanced Chemistry Development |
| ACE | angiotensin-converting enzyme |
| AcrB | multidrug efflux pump protein |
| ADAM | a docking method from the Institute of Medicinal Molecular Design |
| ADME | absorption, distribution, metabolization, excretion |
| ADMET | absorption, distribution, metabolization, excretion, and toxicity |
| AI | artificial intelligence |
| AMBER | Assisted Model Building and Energy Refinement |
| AMOEBA | a force field for proteins |
| API | applications programming interface |
| ASP | atomic solvation parameter |
| AZT | azidothymidine |
| BBB | blood–brain barrier |
| BCI | Barnard Chemical Information |
| C++ | a computer programming language |

| | |
|---|---|
| CAESA | Computer Assisted Estimation of Synthetic Accessibility |
| CAMEO | Computer Assisted Mechanistic Evaluation of Organic Reactions |
| CAOS | computer-aided organic synthesis |
| CASINO | Computer-Aided Synthesis Inference for Organic Compounds |
| CASP | Critical Assessment of Techniques for Protein Structure Prediction |
| CATH | Class, Architecture, Topology, Homology |
| CC | coupled cluster |
| CFD | computational fluid dynamics |
| CFF | Consistent Force Field |
| CFF93 | Consistent Force Field 1993 |
| CFR | Code of Federal Regulations |
| CHARMM | Chemistry at Harvard Macromolecular Mechanics |
| CHEAT | Carbohydrate Hydroxyls represented by Extended Atoms |
| CHIRON | Chiral Synthon |
| CI | configuration interaction |
| CLIX | a docking method |
| ClogP | method for predicting log $P$ |
| CMC | Comprehensive Medicinal Chemistry |
| CNS | central nervous system |
| CoMFA | Comparative Molecular Field Analysis |
| CoMSIA | Comparative Molecular Shape Indices Analysis |
| CPE | Chemical Potential Equalization |
| CPU | central processing unit |
| CSI | Carbó Similarity Index |
| CVFF | Consistent Valence Force Field |
| DEREK | Deductive Estimation of Risk from Existing Knowledge |
| DFT | density functional theory |
| DRF90 | Direct Reaction Field 90 |
| DNA | deoxyribonucleic acid |
| EFF | Electron Force Field |
| EROS | Elaboration of Reactions for Organic Synthesis |
| EVB | Empirical Valence Bond |
| FDA | Food and Drug Administration |
| FEP | free energy perturbation |
| FEP-MD | Free Energy Perturbation Molecular Dynamics |
| FLOG | Flexible Ligands Oriented on Grid |
| FRED | Fast Rigid Exhaustive Docking |

| | |
|---|---|
| FSSP | Fold Classification based on Structure–Structure Alignment of Proteins/Families of Structurally Similar Proteins |
| GA | genetic algorithm |
| GB/SA | Generalized Born Solvent Accessible |
| GLUT2 | glucose transporter 2 |
| GPCR | G-protein-coupled receptor |
| GROMACS | Groningen Machine for Chemical Simulations |
| HADDOCK | High Ambiguity Driven Biomolecular Docking |
| HASL | Hypothetical Active Site Lattice |
| hERG | human ether-a-go-go related gene |
| HF | Hartree–Fock |
| HOLOWin | Holosynthon and Windows |
| HOMO | highest occupied molecular orbital |
| hPEPT1 | human intestinal small peptide carrier |
| HQSAR | hologram quantitative structure–activity relationship |
| HTVS | high throughput virtual screening |
| HUPO | Human Proteome Organisation |
| $IC_{50}$ | concentration at which activity is decreased by 50% |
| ICM | a docking program from MOLSOFT |
| IGOR | Interactive Generation of Organic Reactions |
| InChI | IUPAC International Chemical Identifier |
| IRC | intrinsic reaction coordinate |
| $K_d$ | dissociation constant |
| $K_I$ | inhibition constant |
| $K_M$ | Michaelis constant |
| LBDD | ligand-based drug design |
| $LD_{50}$ | lethal dose for 50% of test subjects |
| LHASA | Logical and Heuristics Applied to Synthetic Analysis |
| LIE | Linear Interaction Energy |
| LIGIN | a docking program |
| $\log D$ | $\log P$ for ionization state at a specific pH |
| $\log P$ | octanol–water partition coefficient |
| $\log S$ | aqueous solubility |
| $\log S_w$ | intrinsic water solubility |
| LOO | leave one out |
| LR | linear regression |
| LUDI | a scoring method for docking and *de novo* design |
| LUMO | lowest unoccupied molecular orbital |
| MAb | monoclonal antibody |
| MCASE | Multi-Computer Automated Structure Evaluation |
| MCS | maximal common subgraph |

| | |
|---|---|
| MD | molecular dynamics |
| MEP | Molecular Electrostatic Potential |
| MFA | Molecular Field Analysis |
| MlogP | a method for predicting log $P$ |
| MLP | molecular lipophilic potential |
| MM | molecular mechanics |
| MM+ | a molecular mechanics force field |
| MM1 | a molecular mechanics force field |
| MM2 | a molecular mechanics force field |
| MM2X | a molecular mechanics force field |
| MM3 | a molecular mechanics force field |
| MM4 | a molecular mechanics force field |
| MMFF | Merck Molecular Force Field |
| MMX | a molecular mechanics force field |
| MOGA | Multiobjective Genetic Algorithm |
| MOMEC | Molecular Mechanics |
| MP$n$ | Møller–Plesset Perturbation Theory ($n = 2, 3, \ldots$) |
| MRSA | methicillin-resistant *Staphylococcus aureus* |
| MSA | molecular shape analysis |
| MVP | Molecular Visualization and Processing Environment |
| NBTI | Non-Boltzmann Thermodynamic Integration |
| NLM | nonlinear map |
| NMR | nuclear magnetic resonance |
| NOE | nuclear Overhauser effect |
| OCSS | Organic Chemistry Synthesis Simulator |
| OPLS | Optimized Potential for Liquid Simulations |
| OPLS-2001 | Optimized Potential for Liquid Simulations 2001 |
| OPLS-2005 | Optimized Potential for Liquid Simulations 2005 |
| OPLS-AA | Optimized Potential for Liquid Simulations All Atom |
| OPLS-UA | Optimized Potential for Liquid Simulations United Atom |
| OSET | Organic Synthesis Exploration Tool |
| OWFEG | One Window Free Energy Grid |
| PAMPA | parallel artificial membrane permeability assay |
| PBE | Poisson–Boltzmann Equation |
| PB/SA | Poisson Boltzmann Solvent Accessible |
| PCA | principal components analysis |
| PFF | Polarizable Force Field |
| $\pi$ (pi) | electron orbitals or bonds perpendicular to the sigma bond |
| p$K_a$ | acidity equilibrium constant |
| PLP | piecewise linear potential |
| PLS | partial least squares |

| | |
|---|---|
| PMF | potential of mean force |
| PTMs | posttranslational modifications |
| QCFF/PI | Quantum Consistent Force Field for Pi electrons |
| QM | quantum mechanics |
| QMFF | Quantum Mechanical Force Field |
| QM/MM | a method combining quantum mechanics and molecular mechanics |
| QPLD | Quantum-Polarized Ligand Docking |
| QSAR | quantitative structure–activity relationship |
| QSM | Quantum Similarity Measure |
| QXP | a force field-based docking program |
| ReaxFF | Reactive Force Field |
| RFF | Reaction Force Field |
| RMSD | root mean square deviation |
| ROCS | Rapid Overlay of Chemical Structures |
| ROSDAL | Representation of Organic Structure Descriptions Arranged Linearly |
| RNA | ribonucleic acid |
| SAR | structure–activity relationship |
| SBDD | structure-based drug design |
| SCR | structurally conserved region |
| SDS | synthesis design systems |
| SCOP | Structural Classification of Proteins |
| SECS | Simulation and Evaluation of Chemical Synthesis |
| SESAM | Search for Starting Materials |
| SIBFA | Sum of Interactions Between Fragments *Ab Initio* Computed |
| SIE | Solvated Interaction Energy |
| SLN | SYBYL Line Notation |
| SMILES | Simplified Molecular Input Line Entry Specification |
| SMoG | Small Molecule Growth |
| SP | standard precision |
| SST | Starting Material Selection Strategies |
| SUA | structural unit analysis |
| SVL | Scientific Vector Language |
| SYBYL | the Greek word for oracle names a force field and software from Tripos |
| SYNGEN | Synthesis Generation |
| TPSA | topological polar surface area |
| UBCFF | Urey–Bradley Consistent Force Field |
| UFF | Universal Force Field |
| VALIDATE | a docking scoring function |

| | |
|---|---|
| VR | variable region |
| WLN | Wiswesser Line Notation |
| WODCA | Workbench for the Organization of Data for Chemical Applications |
| XP | extra precision |
| YETI | a force field |

# BOOK ABSTRACT

Computational techniques play a valuable role in the drug design process. *Computational Drug Design* provides a solid description of those techniques and the roles that they play in the drug design process. This book covers a wide range of computational drug design techniques in an easily understood, nonmathematical format. The emphasis is on understanding how each method works, how accurate it is, when to use it, and when not to use it.

Researchers just learning to do drug design will find this text to be an excellent overview of the entire drug design process. First, the design process is discussed, and then the individual computational techniques are explored in greater depth. Variations on the drug design process for different types of targets are presented. Experienced researchers will be most interested in the tabulations of useful information, such as accuracies of docking methods, and bioisosteres.

*Computational Drug Design* features:

- a discussion of the drug design process and how that process differs depending upon the specific drug target
- chapters covering each of the computational techniques used in the drug design process
- comparisons between specific implementations of each method

## ABSTRACT OF CHAPTERS

### 1. Introduction

- Drug design is a difficult and costly process.

### 2. Properties that Make a Molecule a Good Drug

- Drugs must meet multiple criteria of being active, bioavailable, and non-toxic.

### 3. Target Identification

- A protein in the appropriate metabolic pathway must be found in order to find a way to treat the disease. The three-dimensional structure of that protein must be known in order to use rational drug design techniques to design drugs for it.

### 4. Target Characterization

- The reaction mechanism that the target undergoes must be known in order to determine if a drug could be a competitive inhibitor, allosteric, or follow some other mechanism.

### 5. The Drug Design Process for a Known Protein Target

- Structure-based drug design (also called rational drug design) is the process of designing drug to fit in a particular protein's active site.

### 6. The Drug Design Process for an Unknown Target

- Structure–activity relationships can be used to design a drug, even when the target is not known.

### 7. Drug Design for Other Targets

- Drugs can also be designed to interact with DNA and RNA. There are also differences in the design process for creating steroids and allosteric inhibitors.

### 8. Compound Library Design

- There are software packages to facility the design of lists of compounds to be synthesized through combinatorial synthesis.

9. **Homology Model Building**

    - Homology models are three-dimensional models of proteins designed by aligning to a template structure.

10. **Molecular Mechanics**

    - Molecular mechanics is the molecular energy computation mechanism behind many computational drug design techniques.

11. **Protein Folding**

    - Protein folding is the process of creating a three-dimensional structure for a protein from its primary sequence, without the use of a template structure.

12. **Docking**

    - Docking is the process of determining the orientation and energy of binding of a compound in a protein's active site.

13. **Pharmacophore Models**

    - A pharmacophore is a three-dimensional description of the features needed for activity. This is used to search databases of compounds for appropriate structures.

14. **QSAR**

    - Quantitative structure–activity relationships are used to predict molecular properties such as bioavailability and toxicity.

15. **3D-QSAR**

    - 3D-QSAR is a technique for predicting how a drug will interact with an active site when the geometry of the active site is not known.

16. **Quantum Mechanics in Drug Design**

    - Quantum mechanical techniques are used to determine reaction mechanisms, and to give highly accurate results.

17. ***De novo* and Other AI Techniques**

    - *De novo* programs use artificial intelligence algorithms to automate the structure-based drug design process.

**18. Cheminformatics**

- Cheminformatics refers to the process of storing and searching large quantities of chemical data.

**19. ADMET**

- Adsorption, distribution, metabolization, excretion, and toxicity are properties of molecules that are as important as activity when creating a marketable drug.

**20. Multiobjective Optimization**

- Multiobjective optimization is the process of finding solutions that simultaneously fit multiple criteria.

**21. Automation of Tasks**

- Drug design processes can be automated to handle large numbers of compounds through the use of scripting languages, icon-oriented programming systems, and spreadsheet-based programs.

**22. Bioinformatics**

- Bioinformatics techniques are used for the analysis of DNA and protein sequences.

**23. Simulations at the Cellular and Organ Level**

- Cells, membranes, and organs can also be simulated.

**24. Synthesis Route Prediction**

- There are computational tools for suggesting synthetic strategies.

**25. Proteomics**

- Proteomics is the study of the proteome.

**26. Prodrug Approaches**

- Prodrugs are compounds that are converted to active drugs after entering the body.

## 27. Future Developments in Drug Design

- There are a number of potential future developments in the pharmaceutical industry, including stem cells, cloning, genetic manipulation, and increasing longevity.

# CHAPTER KEYWORDS

## 1. Introduction

- cost
- difficulty

## 2. Properties that Make a Molecule a Good Drug

- assay
- drug-likeness

## 3. Target Identification

- metabolic pathway
- crystallography

## 4. Target Characterization

- mechanism
- reaction coordinate
- transition structure
- molecular dynamics

## 5. The Drug Design Process for a Known Protein Target

- structure-based drug design
- compound refinement
- resistance

## 6. The Drug Design Process for an Unknown Target

- ligand-based drug design

## 7. Drug Design for Other Targets

- DNA
- RNA

- allosteric
- steroids
- central nervous system

**8. Compound Library Design**

- targeted library
- diverse library
- enumerative
- non-enumerative

**9. Homology Model Building**

- template
- alignment

**10. Molecular Mechanics**

- force field
- Monte Carlo

**11. Protein Folding**

- protein folding
- conformational analysis

**12. Docking**

- docking
- scoring
- flexible active site
- hierarchical docking

**13. Pharmacophore Models**

- pharmacophore

**14. QSAR**

- QSAR
- descriptors

## 15. 3D-QSAR

- 3D-QSAR
- CoMFA

## 16. Quantum Mechanics in Drug Design

- accuracy
- reaction path

## 17. *De novo* and Other AI Techniques

- *de novo*
- artificial intelligence

## 18. Cheminformatics

- cheminformatics
- similarity
- substructure search
- clustering

## 19. ADMET

- toxicity
- adsorption
- metabolism
- elimination

## 20. Multiobjective Optimization

- multiobjective optimization

## 21. Automation of Tasks

- script language

## 22. Bioinformatics

- bioinformatics

## 23. Simulations at the Cellular and Organ Level

- cellular simulation
- organ simulation

## 24. Synthesis Route Prediction

- CAOS

## 25. Proteomics

- proteomics
- proteome

## 26. Prodrug Approaches

- prodrug

## 27. Future Developments in Drug Design

- genetic manipulation
- cloning
- stem cells
- longevity

# 1

# INTRODUCTION

## 1.1 A DIFFICULT PROBLEM

Let us take an incredibly simplified view of the statistics of drug design. There are an estimated 35,000 open reading frames in the human genome, which in turn generate an estimated 500,000 proteins in the human proteome. About 10,000 of those proteins have been characterized crystallographically. In the simplest terms, that means that there are 490,000 unknowns that may potentially foil any scientific effort.

The previous paragraph is far from being a rigorous analysis. However, it does illustrate the fact that drug design is a very difficult task. A pharmaceutical company may have from 10 to 100 researchers working on a drug design project, which may take from 2 to 10 years to get to the point of starting animal and clinical trials. Even with every scientific resource available, the most successful pharmaceutical companies have only one project in ten succeed in bringing a drug to market.

Drug design projects can fail for a myriad of reasons. Some projects never even get started because there are not adequate assays or animal models to test for proper functioning of candidate compounds. Some diseases are so rare that the cost of a development effort would never be covered by product sales. Even when the market exists, and assays exist, every method available may fail to yield compounds with sufficiently high activity. Compounds that are active against the disease may be too toxic, not bioavailable, or too costly to manufacture. In all fairness, we should note that high manufacturing cost is seldom a sufficient deterrent in the pharmaceutical industry.

Sometimes the only compounds that work are already the competitor's intellectual property. This book will not be addressing intellectual property law, but we shall point out the following thumb rule of commercial product development:

> *A product does not have to be better than the competitor's product. It has to be about as good as the competitor's product and patentable under your own name.*

Biological systems are probably one of the most complex systems under study on the planet. Not surprisingly, drugs are seldom simple molecules. Most are heterocyclic, are of moderate molecular weight, and contain multiple functional groups. As such, the challenges of organic synthesis are sometimes as great as the challenge of determining what compounds should be synthesized. In the pharmaceutical industry, the answer is often to synthesize all possible derivatives within a given family of compounds.

In the course of computational drug design, researchers will find themselves tasked with solving a whole range of difficult problems, including efficacy, activity, toxicity, bioavailability, and even intellectual property. With the total drug development process costing hundreds of millions of dollars, and enormous amounts of money being spent daily, drug design chemists can be under incredible pressure to produce results. As such, it is necessary to effectively leverage every computational tool that can help to achieve successful results. This book has been written to give a solid understanding of the whole range of available computational drug design tools.

## 1.2 AN EXPENSIVE PROBLEM

There have been a number of published estimates of how much it costs to bring a drug to market. Recent estimates have ranged from $300 million to $1.7 billion. A single laboratory researcher's salary, benefits, laboratory equipment, chemicals, and supplies can cost in the range of $200,000 to $300,000 per year. Some typical costs for various types of experiments are listed in Table 1.1.

Owing to the enormous costs involved, the development of drugs is primarily undertaken by pharmaceutical companies. Indeed, the dilution of investment risk over multiple drug design projects pushes pharmaceutical companies to undertake many mergers in order to form massive corporations.

**TABLE 1.1 Typical Costs of Experiments**

| Experiment | Typical Cost per Compound ($) |
|---|---|
| Computer modeling | 10 |
| Biochemical assay | 400 |
| Cell culture assay | 4,000 |
| Rat acute toxicity | 12,000 |
| Protein crystal structure | 100,000 |
| Animal efficacy trial | 300,000 |
| Rat 2-year chronic oral toxicity | 800,000 |
| Human clinical trial | 500,000,000 |

Only rarely are drugs taken all the way through the approval process by academic institutions, individuals, government laboratories, or even small companies. In 1992, out of the 100 most prescribed drugs, 99 were patented by the pharmaceutical industry.

## 1.3 WHERE COMPUTATIONAL TECHNIQUES ARE USED

There is no one best computational drug design technique. Many techniques are used at various stages of the drug design project. At the beginning of a project, cheminformatics techniques are used to select compounds from available sources to be assayed. Once some marginally active compounds are found, relatively broad similarity searching techniques are used to find more compounds that should be assayed. As larger collections of more active compounds are identified, the computational chemists will shift to successively more detailed techniques, such as QSAR, pharmacophore searching, and structure-based drug design tools such as docking. A computational chemist may make their reputation by being a world-class expert at the design or use of one of these techniques. However, a functional knowledge of how to work with many of them is usually necessary in order to be successful as a computational chemist in the pharmaceutical industry.

The simplest form of drug design is to start with a marginally active compound, and then make slightly modified derivatives with slightly different functional groups. However, this type of trial-and-error modification of molecules is a "blind man's bluff" game, until you see how those molecules fit in the active site and interact with the protein residues. Thus, the majority of the time that researchers are designing structures "by hand" today, they do so by examining the way that the compounds fit in the target's active site as displayed through three-dimensional computer rendering. Once a compound has been built within such computer programs, it is easy to subsequently test how strongly it will bind in the active site using computational techniques such as docking.

Computational techniques provide other options for understanding chemical systems, which yield information that is difficult, if not nearly impossible, to obtain in laboratory analysis. For example, quantum mechanically computed reaction coordinates can show the three-dimensional orientation that species adopt at each step of a reaction mechanism. Likewise, they can show exactly where the unpaired spin density is located at each point along a reaction coordinate. This is of particular concern in drug design, since enzymes often catalyze reactions by holding species in the preferred orientation, and sometimes include a mechanism to provide for necessary electron or hydrogen transfer.

Computer simulations are less costly per compound than any laboratory test, as illustrated in Table 1.1. Because of this cost-efficiency, large databases of compounds are often tested in software. Many of these compounds will never see laboratory testing of any sort. Indeed, many compounds are designed and tested in software, but never synthesized at all, owing to poor results *in silico* (computer calculations). Likewise, the use of computational techniques to choose compounds for testing results in an enrichment, meaning that a higher percentage of the compounds that are tested are active.

In today's world of mass synthesis and screening, the old practice of sitting down to stare at all of the chemical structures on a single sheet of paper is hopeless. Drug design projects often entail having data on tens of thousands of compounds, and sometimes hundreds of thousands. Computer software is the ideal means for sorting, analyzing, and finding correlations in all of this data. This has become so common that a whole set of tools and techniques for handling large amounts of chemical data have been collectively given the name "cheminformatics."

The problems associated with handling large amounts of data are multiplied by the fact that drug design is a very multidimensional task. It is not good enough to have a compound that has the desired drug activity. The compound must also be orally bioavailable, nontoxic, patentable, and have a sufficiently long half-life in the bloodstream. The cost of manufacturing a compound may also be a concern—less so for human pharmaceutics, more so for veterinary drugs, and an extremely important criterion for agrochemicals, which are designed with similar techniques. There are computer programs for aiding in this type of multidimensional analysis, optimization, and selection.

Most importantly, drug design projects may fail without the efforts of experts in computational modeling. Drug design is such a difficult problem that every relevant technique is often utilized to its best advantage. Computational modeling techniques have a long history of providing useful insights, new suggestions for molecular structures to synthesize, and cost-effective (virtual) experimental analysis prior to synthesis.

Thus, computational drug design techniques play a valuable role in pharmaceutical research. This role makes computational techniques an important part of a successful and profitable drug design process.

## BIBLIOGRAPHY

### Cost of Drug Development

DiMasi JA, Hansen RW, Grabowski HG. The price of innovation: New estimates of drug development costs. J Health Econ 2003; 22: 151–185.

Gilbert J, Henske P, Singh A. Rebuilding Big Pharma's business model. In Vivo: The Business & Medicine Report 2003; 21(10).

Klaassen CD. Principles of toxicity. In: Klaassen CD, Amdur MO, Doull J, eds. The Basic Science of Poisons. Columbus, OH: McGraw-Hill; 1986. p. 11–32.

Tufts Center for the Study of Drug Development. Total cost to develop a new prescription drug, including cost of post-approval research, is $897 million. Available at http://csdd.tufts.edu/NewsEvents/RecentNews.asp?newsid=29. Accessed 2006 Apr 13.

### Computational Drug Design Process in General

Boyd DB. Drug design. In: von Ragué Schleyer P et al., editors. Encyclopedia of Computational Chemistry. New York: Wiley; 1998. p. 795–804.

Additional references are contained on the accompanying CD.

# PART I

# THE DRUG DESIGN PROCESS

# 2

# PROPERTIES THAT MAKE A MOLECULE A GOOD DRUG

Once when doing usability testing on a piece of drug design software, we asked an open-ended question to see what most interested chemists. The question asked was simply "Which of these compounds do you find interesting?" We had hoped that this type of question would be open enough for the chemists to start asking for new features, such as a chemically relevant way to sort the molecules. Instead, the question told us more about the person we asked it of than about the software. If the subject was trained as a synthetic organic chemist, they chose a molecule that they could easily synthesize. If the subject was trained in computational drug design, they chose the most drug-like molecules, those that were heterocyclic with multiple functional groups, and few functional groups known to be highly toxic.

Obviously, both being able to synthesize a molecule and choosing to pursue synthesis of compounds that could potentially be useful drugs are important concerns. This book is not about organic synthesis, although Chapter 24 discusses some software packages that have been created to help map out possible synthesis routes. This chapter is devoted to discussing why some compounds are considered to be "drug-like" and others are not. The first section talks about the ways in which compounds are tested for usefulness as a drug. Discussing why compounds pass or fail these tests will begin to give some nonspecific insight into what features of molecular structure are important in drug design. The second section is devoted to discussing properties of molecules

*Computational Drug Design*. By David C. Young

that determine whether they have the potential to be good drugs. Finally, we present some exceptions to the typical rules of drug-likeness.

## 2.1 COMPOUND TESTING

"Efficacy" is the qualitative property of a compound having the desired effect on a biological system. In the case of drug efficacy, this means having a measurable ability to treat the cause or symptoms of a disease. "Activity" is the quantitative measure of how much of that compound is required to have a measured effect on the biological system. Drugs work through binding to a target in the body, or a pathogenic organism such as a virus or bacterium. The target is usually a protein, but in some cases can be DNA, RNA, or another biomolecule. The vast majority of drugs work by inhibiting the action of the target. Unless explicitly stated otherwise (see Chapter 7), it is assumed in this book that drug activity is obtained through inhibition of the target.

It is important to understand the terminology related to drug testing. As compounds show various levels of activity in the different stages of testing, the compounds will be referred to as "hits," "leads," "drug candidates," "drugs," and several other terms. Within each pharmaceutical company, these words have very precisely defined meanings. However, the technical definitions of these terms differ from one company to another. Sometimes, even within the same company, the terminology used within the research and development laboratories will be different from that used by the marketing department, or there may be differences in the quantitative criteria from one project to the next. Thus, researchers must be careful about how their company defines these terms, and to understand the difference in terminology when talking with researchers from other companies.

Typically, the term "hit" refers to compounds identified in some initial rounds of screening. Compounds identified as hits typically go through additional rounds of screening. Once screening results have been verified, some readily obtained derivatives will be synthesized or purchased and tested. Once a compound, or often a series of compounds, meets a certain set of criteria, the series will be designated as a lead series. The lead series is then used as the basis for a more comprehensive synthesis of many derivatives and more in-depth analysis, both computational and experimental. The following are some typical criteria that may be necessary to move a compound series onto the lead development stage:

- concentration-dependent activity
- active in both biochemical and cell-based assay
- below some $IC_{50}$ threshold (perhaps low micromolar, or down to the nanomolar range)

- some understanding of the series' structure–activity relationships
- known binding kinetics
- selectivity assessment
- well-established structure and purity
- stability assessment
- synthetically tractable
- series is patentable
- some path for optimization (creating derivatives) is apparent
- solubility measured
- log *D* measured
- metabolic liabilities predicted
- pharmacokinetics predicted or measured
- toxicity issues predicted
- potential for significant side effects considered (e.g., whether the drug will block hERG channels, thus resulting in drug-induced cardiac arrhythmia)

Since the precise definitions of terms such as "hit" and "lead" are not universally accepted, compounds at any stage of testing will be referred to in this book using terms that are generically interchangeable within the context of the book, such as "compound," "molecule," or "drug."

In order to test if a compound is effective and to measure its activity, four different types of experimental tests may be used:

- biochemical assays
- cell-based assays
- animal tests
- human tests

For each of these types of tests, the results will give different information about how the compound interacts with the biological system.

The following sections give a short discussion of each type of test. Note that crystallography is not mentioned here, since it is used more for target development than for testing of drug efficacy and is thus addressed in the appropriate section of this book. Most of the testing procedures and their details are beyond the scope of this text. The following discussion is intended only to give an introductory description and to establish some terminology. Later in the book, there will be discussions of how these tests fit in with the computational drug design process.

### 2.1.1 Biochemical Assays

Biochemical assays are usually the cheapest way of experimentally testing compounds. They are also the simplest environment in which the compound can be tested. Thus, biochemical assays are usually the first line of experimental testing of compounds. The term "*in vitro* assay" is sometimes applied to biochemical assays or cell-based assays, or both.

Biochemical assays entail putting the drug target, usually a protein, in solution with a compound, and measuring whether the compound inhibits the protein's activity. Ideally, this can be done using very small quantities of material in the wells of a plate. If possible, it is preferable to measure results colorimetrically. The quantitative measurement of activity is referred to as "scoring" the plates.

In high-throughput screening, there might be a single test for a given compound, in a single well of a plate. This can only give a yes/no result at some activity threshold. Single-well tests are prone to having a percentage of false-positive and false-negative results, due to random fluctuations in reaction conditions and other factors. These tests are usually used to perform a first-pass screening of large libraries of compounds. Typically, libraries that represent a very wide range of molecular structures are screened in this way. This is done in order to generate an initial list of compounds that may be active. Diverse screenings sometimes identify a class of compounds that inhibit a given target and have not been explored in the past for activity against that target.

Promising compounds are usually retested at a series of different concentrations. This allows a value of the inhibition constant $K_I$ to be computed from the results. Sometimes, assay results are reported as an $IC_{50}$ value, which is the concentration at which the target activity is decreased by 50%.

A passed biochemical assay should indicate that the compound has bound to the target's active site, thus having activity (inhibiting the operation of the target) through a competitive inhibition mechanism. However, it is possible to have false-positive results. For example, if the results are being measured colorimetrically, then colored compounds can give false-positive results. There are sometimes ways to avoid this problem by careful selection of the measuring wavelengths or by subtracting out the signal from a blank solution.

Compounds that show negative results in a biochemical assay are typically those that do not bind to the target. However, there can be false-negative results. For example, these failures can be due to the compound being tested not being soluble enough to stay in solution.

In general, biochemical assay results are the experimental tests that show the highest correlation with computational drug design techniques. Since techniques such as docking, 3D-QSAR, and pharmacophore searches model the compound in the active site of the target, there is a direct correspondence between the computational results and biochemical assay results. Often, this correspondence is used in both directions. Biochemical assay results are used to verify that appropriate computational techniques have been chosen for the target being studied. Then computational results are used to flag when an experimental result should be examined more closely as a possible false-negative or false-positive result. Once validated (preferably), the computational techniques can be used to search for potentially active compounds in massive libraries of chemical structures at a far lower cost than experimentally testing each compound.

### 2.1.2 Cell-Based Assays

Cell-based assays (sometimes called *in vitro* assays) are performed by placing a compound in solution with a culture of living cells. This adds a level of complexity to the test conditions. Tests for a compound inhibiting an enzyme inside the cell may fail because of a lack of efficacy or because the compound is not sufficiently lipophilic to pass through the cell membrane. Several types of cell culture assays are often performed. Compounds showing efficacy may also be tested for cytotoxicity and mutagenicity using cell-based assays.

Cell-based assays are typically more expensive than biochemical assays, but still fill a valuable role. The greater cost is mostly due to their being more labor-intensive and more difficult to automate. However, there are some assays, such as ion channel assays, that cannot be done biochemically and therefore must use living cells. The cell-based assay results give additional information, such as bioavailability, to decide which compounds should go to an initial synthesis scale-up and into animal testing.

If a cell-based assay indicates that a compound is active, it is an indication that activity should be seen if the drug gets into the blood plasma. This is not always true, since it does not address oral bioavailability, blood–brain barrier permeability, reaching the correct organ, etc. It does at least give an indication of correct binding to cell surface receptors, or that the drug is reaching a target inside a cell.

A failed cell-based assay can mean that the drug is not active against the target, but there are also other possible reasons for failures. A failure can mean that the compound is cytotoxic or mutagenic, which can be a disqualifying criterion in itself. There could also be a bioavailability issue, such as not reaching a target internal to a cell. Or the compound may be insoluble under the test conditions.

Another type of cell-based assay is a Caco-2 assay. Caco-2 cells are colon wall carcinoma cells. There is a Caco-2 cell line that can grow a layer of cells that aligns to have the efflux pumps (P-glycoprotein and less common ones) arranged to provide active transport of chemical species from one side to another. Two tests can be performed to see how much of a given species will be transferred in each direction. By comparing these two results, it is possible to get an indication of how much the species will be affected by both active and passive transport.

The advantage of the Caco-2 assay is that it is one of the few experimental bioavailability measurements available. Researchers are really interested in passive absorption in the small intestine, where most substances enter the body, and in diffusion through the cell membrane. However, since assays for those are either not available or much more difficult to run, Caco-2 assays are performed with the understanding that these other types of bioavailability are generally proportional to the Caco-2 assay results. An alternative to Caco-2 is a MDR1-MDCK cell line assay, which is based on dog kidney cells. Another alternative is a parallel artificial membrane permeability assay (PAMPA).

### 2.1.3 Animal Testing

Animal testing (also called *in vivo* testing) is much more expensive than biochemical and cell culture testing. Thus, only a handful of compounds will be sent on to animal testing. Animal testing typically provides an indication of whether the compound shows efficacy in a mammal, is toxic, and is orally bioavailable. Animal testing can also give an initial indication of whether a potential drug displays severe side effects. Animal testing typically takes about 6.5 years if testing for toxicity with chronic dosages is performed. Animal efficacy testing and acute dosage toxicity tests are much less time-consuming.

Ideally, animal testing should be done with mice, for cost reasons. However, a necessary prerequisite for efficacy testing is having an animal susceptible to the disease being treated. Thus, it is sometimes necessary to do animal testing with rats, rabbits, dogs, primates, armadillos, or other species. For example, the only animal model for leprosy is the armadillo.

An increasingly important tool is the creation of knockout mice. These are mice that have one gene turned off through genetic manipulation. This is not always possible, as some gene deactivations cause prenatal or postnatal mortality of the mice. Studying the effect of a gene knockout can give information about the role that that gene plays in the disease under study. It also gives indications as to how downregulating the function of that gene through the action of a drug will affect the animal.

Another option is the creation of transgenic mice. These are mice that have an extra gene, not found in mice, inserted into their genome. If there is not an animal available that is susceptible to a given disease, it may be possible to create transgenic mice that are susceptible. For example, transgenic mice susceptible to polio, which normal mice do not contract, have been created. There have also been transgenic sheep and chickens created to manufacture human proteins. At present, the creation of knockout mice and especially transgenic animals is still a very difficult and expensive task.

In recent years, there have been a number of attempts to find alternatives to animal testing. In some cases, there are viable alternatives, such as testing on animal tissues. Where alternatives exist, they are often used both for cost savings reasons and for humane reasons. However, there are still many diseases for which there is no alternative to animal testing that is acceptable to the United States Food and Drug Administration (FDA).

### 2.1.4 Human Clinical Trials

For every drug, there must be a first time that it is tried on a human being. This necessarily entails some risk to human life. Human clinical trials have been designed as a multiple-step process, referred to as phases, in order to minimize the risk of harming patients while doing the testing necessary to bring the drug to market as quickly as possible. The following discussion is based on the clinical trial regime used in the United States, which is regulated by the FDA. The testing requirements vary greatly from one country to another. Many countries have an abbreviated testing process for drugs that have already been approved in another country, particularly if the drug has been approved in the United States, which has some of the most stringent testing requirements in the world.

***2.1.4.1 Phase I Clinical Trials.*** The primary goal of phase I clinical trials is safety testing. The test is designed to verify that the drug does not result in any severe side effects that would be harmful to a healthy adult. Thus, the subjects in a phase I trial do not have the disease that the drug was designed to treat. Often phase I trial subjects are college students, who are paid a small sum of money to participate in trials being carried out at university-affiliated facilities. Phase I clinical trials typically take about a year and a half. About 70% of the drugs submitted to phase I testing are sent on to phase II testing.

***2.1.4.2 Phase II Clinical Trials.*** Phase II trials are designed to obtain an initial indication of whether the drug successfully treats the disease in humans. They also give some initial data about necessary dosages and side effects. Typically, a few hundred patients at various stages of disease progression are tested in phase II trials. Phase II typically takes about two years. About a third of the drugs tested in phase II trials are sent on to phase III testing.

***2.1.4.3 Phase III Clinical Trials.*** Phase III trials entail giving the drug to a much larger group of patients. In this phase, dosage regimes are determined more precisely. There will also be a sufficiently diverse sample of patients to determine if the necessary dosage is affected by other drugs that the patients are taking or by patient ethnic background. Patients are also monitored for side effects, both severe and mild. Typically, a few thousand patients are treated in a phase III trial. About a quarter of the compounds submitted to phase III trials are approved for use in the United States.

After phase III testing is completed, the results are submitted to the FDA for approval as a prescription medication. Data on patient histories continues to be collected long after FDA approval. This data is later used to justify approval for sales over the counter (without a prescription). Yet, later in the life of a drug, the patents expire, making it fair game for generic manufacturers.

There is a considerable amount of variation in clinical testing from one drug to another. If a drug is administered to patients with only months left to live, it may be possible to very quickly show that the drug extends life with an acceptable level of side effects. Drugs that are taken repeatedly over a patient's life, such as cold or allergy medications, require very extensive testing, and may be approved only if side effects are extremely mild.

Most compounds do not fail clinical trials owing to failing to treat the disease. Most fail owing to ADMET issues ("absorption, distribution, metabolization, excretion, and toxicity"). Some are eliminated from the body so quickly that the patient must take doses of the medicine every few hours. Some have unacceptable side effects. Some have poor oral bioavailability, so massive quantities of pills must be taken in order to get enough into the bloodstream. Designing compounds that address all of these issues as well as treating an illness is one of the reasons that drug design is such a challenging task. The first step is to look at how various aspects of molecular structure correlate with drug activity, bioavailability, and toxicity.

## 2.2 MOLECULAR STRUCTURE

The issues discussed above can be translated into constraints on molecular structure. The following is a discussion of why certain molecular structures might be desirable or undesirable as drugs. Although many generalizations can be made, some of these depend upon the particular drug target being studied, and thus there are exceptions to all of these loose rules. The last section of this chapter will examine some exceptions to the rules.

### 2.2.1 Activity

The first concern of drug designers is to create compounds that will have very high activity. Activity is usually quantified by the order of magnitude of the biochemical assay inhibition constant $K_I$. Nanomolar activity is usually the objective of drug design efforts. In rare cases, picomolar activity can be achieved. It is very difficult to start from a compound with millimolar activity and modify the molecular structure to obtain a compound with nanomolar activity. However, that is usually the situation that drug designers face. It is considerably easier to start with a compound having subnanomolar activity and modify it to have better bioavailability or less toxicity without sacrificing too much of the activity.

The majority of drug targets are proteins, such as enzymes or cell surface receptors. In the majority of cases, the drug acts through competitive inhibition, by binding to the active site and thus preventing the native substrate from entering the site. Thus, the first major requirement is that the drug should be the correct shape to fit in the active site. This is referred to as the lock-and-key theory of drug action.

The second requirement is that the drug should bind to the active site. This is accomplished through the positioning of functional groups in the drug molecule. For example, if the active site contains a hydrogen bond donor, then the drug should have a hydrogen bond acceptor positioned to give a hydrogen bond binding the drug to the active site. Other important interactions are $\pi$-system stacking, positioning of charged groups to form ionic bonds, van der Waals interactions, and steric hindrance. In a minority of cases, discussed later in this book, the inhibitor will form a covalent bond with the target's active site.

Many drugs resemble the target's native substrate; such drugs are referred to as substrate analogs. For example, if the native substrate is a protein sequence including a proline, then the proline ring could be mimicked by a tetrahydrofuran ring. Understandably, a compound that is very similar to the native substrate in shape and binding properties will also tend to be accepted into the active site. Typically, a good drug will be sufficiently different from the native substrate that it will not react with the target, or be too readily metabolized by the body.

One very successful tactic is to create drugs that are transition state analogs. In the case of an enzyme that catalyzes a reaction, the reaction can be described as an initial complex of reactants, a transition structure, and a complex of products. A transition state analog resembles the transition structure for the reaction, but lacks the necessary moiety (i.e., a leaving group) to allow the reaction to go to completion. The enzyme will accept the drug, form a complex that resembles the transition structure, and then get stuck at that point unable to complete the reaction. For example, the first step in the reaction by which proteases cleave amide bonds is a nucleophilic attack on the carbonyl carbon in the –N–CO– group. If the drug resembles a modified version of the native substrate in which that amide is replaced by –C–CO–, $-N-PO_2-$, or $-O-PO_2-$, it will still undergo nucleophilic attack, but the protease will be unable to cleave the bond. Some authors treat the terms "substrate analog" and "transition state analog" as synonymous, and others make the mechanistic distinction described here.

Experimentally, the positioning of the drug in the active site can be seen only with great effort by generating crystal structures of the protein with the drug soaked into the crystal. The experimental analysis of the binding is most often obtained by measuring an inhibition constant $K_I$ with a biochemical assay. Both fit and binding energy are much easier to analyze computationally. The primary tool for this analysis is the "docking" technique, in which an automated algorithm positions the molecule in many different orientations in the active site to find the lowest-energy orientation. This direct correlation between computational results and drug activity makes docking the mainstay of structure-based drug design. Pharmacophore models and 3D-QSAR also model binding and fit to the active site, although somewhat more indirectly. For more information on docking, see Chapter 12. The majority of drugs bind in a crevice in the protein surface as shown in Fig. 2.1.

Binding to an active site is important, but it is not the only criterion to be considered. Another important issue is that of specificity. Specificity is the ability of a compound to bind to the target protein, but not to structurally similar proteins. Every time a drug binds to the wrong protein, it is a possibility

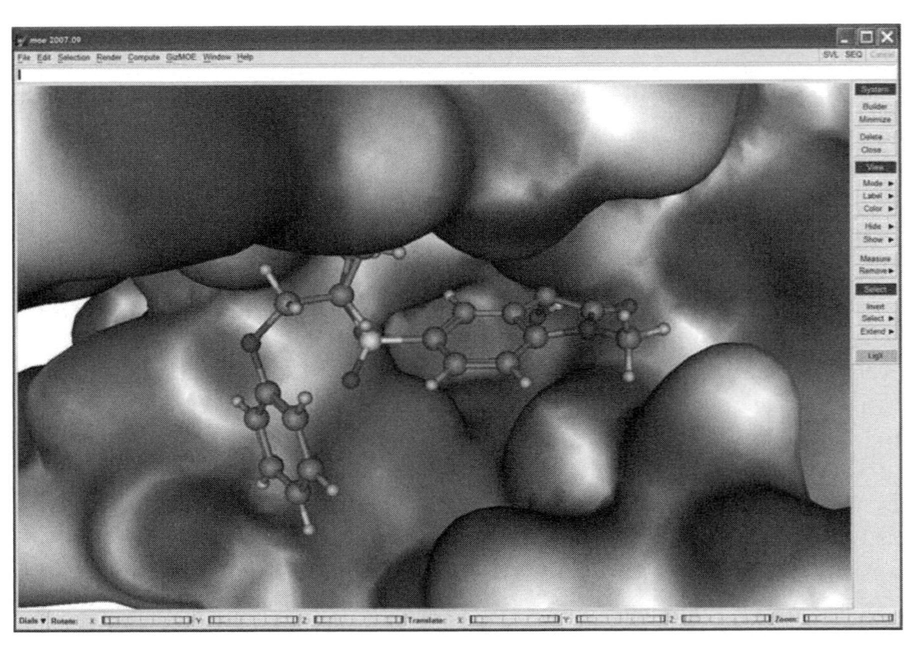

**Figure 2.1** An isatin sulfonamide inhibitor in the caspase-3 protein active site crevice (pdb 1GFW). This is displayed within the MOE software package from Chemical Computing Group, a full-featured drug design package with a rich variety of options for displaying the system. The colors on the surface indicate areas with negative, positive, or neutral partial charges on the residues. A color copy of this figure is contained on the accompanying CD.

for the drug to display unwanted side effects when administered to patients. Protein structural similarity data is given by a systematic Structural Classification of Proteins (SCOP), which is cataloged online at http://scop.mrc-lmb.cam.ac.uk/scop. Other useful protein comparison systems are CATH (Class, Architecture, Topology, and Homology) at http://www.cathdb.info, and FSSP (Families of Structurally Similar Proteins) at ftp://ftp.ebi.ac.uk/pub/databases/fssp/fssp.html and http://srs.ebi.ac.uk/srsbin/cgi-bin/wgetz?-page+LibInfo+-lib+FSSP. Typically, docking studies are used to test how strongly compounds bind in a list of proteins that have active sites very similar to that of the target protein. A duplicate set of biochemical assays is often performed for just a couple of the proteins having the most structurally similar active sites.

Some protein active sites are described as having specificity pockets. These are empty areas of the active site, which may not be particularly important for drug binding. However, designing a drug to have a nonpolar functional group oriented in the specificity pocket can improve the drug's specificity. The drug's ability to bind in the active site of similar proteins will be diminished by the steric hindrance of this additional functional group. An example of a specificity pocket is presented in Fig. 2.2.

It is possible to design a drug with too much specificity. This is of particular importance when designing compounds to inhibit viral proteins. If the compound fits too tightly in an active site, a minor mutation of the virus can change the active site slightly, thus preventing the drug from binding. This is one mechanism by which viruses develop resistance to drugs. When designing antiviral drugs, it is important to test and design against the protein structure from several serotypes of the virus.

It is important to remember that the shape of the active site may not stay fixed. There are some proteins in which the geometry of the active site changes depending upon what is in the site. This is referred to as having an induced fit. It is estimated that as many as 50% of proteins operate in this way. In some cases, individual residues in the active site will have a functional group that undergoes a conformational change to swing into or out of the active site. In other cases, an entire loop may change conformation to fold down on top of the active site. Sometimes, two halves of the protein move in a clamshell-like motion. A few computational drug design tools are designed to automatically include active site flexibility in the simulation. Many do not, thus making it the researcher's responsibility to run multiple calculations with likely active site conformations. An example of an induced fit mechanism is presented in Fig. 2.3.

Difficulties can be encountered when attempting to use a molecule with a large number of rotatable bonds as a drug. The active conformation may not be the lowest-energy conformation. The energy penalty of adopting the active conformation results in a net decrease in the binding energy. A

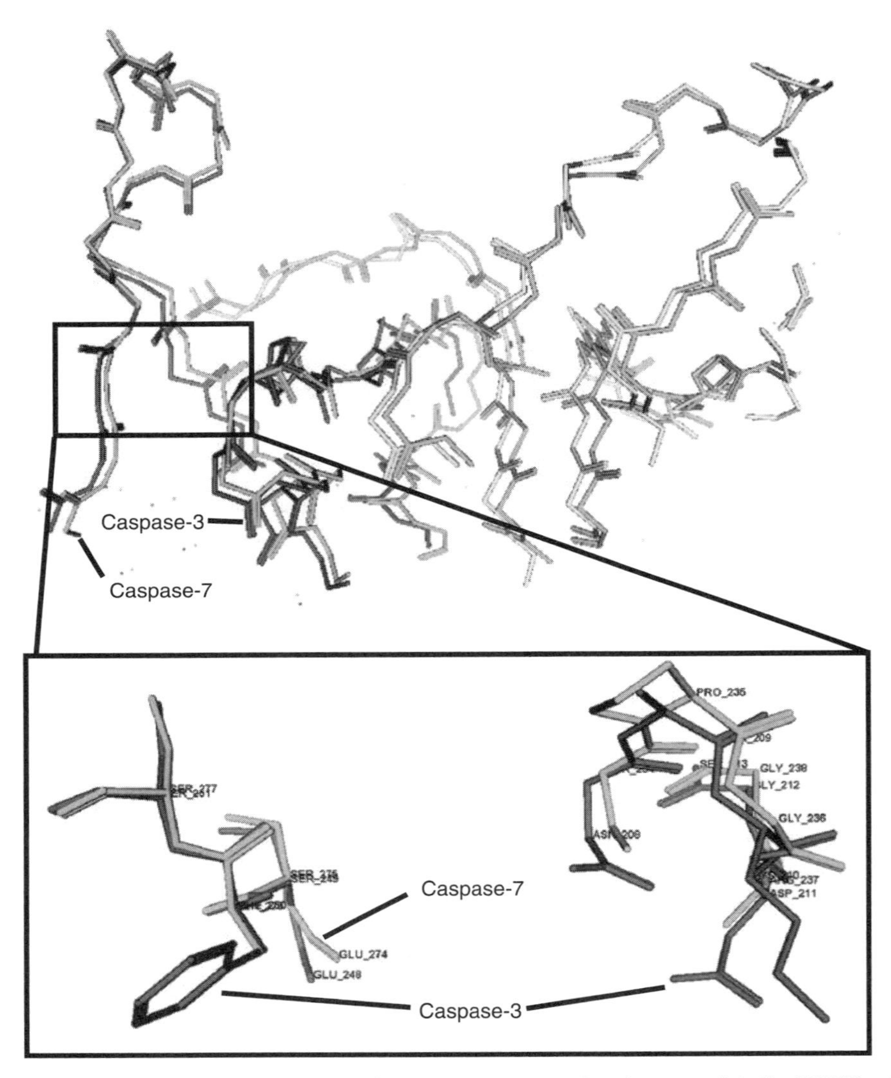


**Figure 2.2** Specificity pocket comparison between caspase-3 and caspase-7 (pdbs 1GFW and 1I40). The backbone alignment near the active site (top) shows how similar these two enzymes are. The close-up view with the side chains at the lower end of the active site (bottom) shows where drug interactions must occur in order to create an inhibitor specific to one but not the other.

conformationally flexible molecule may also be able to adopt a conformation that would bind in the active sites of similar proteins, thus giving specificity problems. Finally, the binding kinetics can be harmed by failed binding events in which the molecule gets to the active site, but does not bind owing to being in the wrong conformation. In light of these potential problems, it is understandable that a significant percentage of approved drugs have either

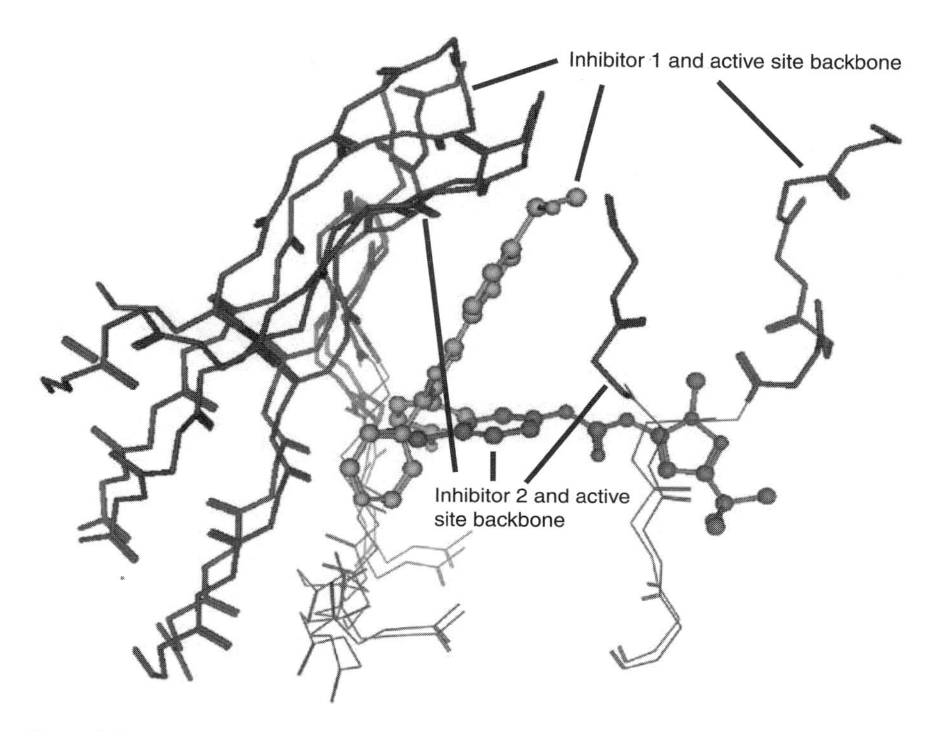


**Figure 2.3** The p38 MAP kinase active site exhibits an induced fit. Here it is shown in two different conformations, induced by two different inhibitors (pdbs 1A9U and 1KV1).

fused ring systems or double bonds. More rigid molecules can sometimes give better activity and fewer side effects.

One potential trap to avoid in drug design is the use of promiscuous binders (also called frequent hitters or privileged structures). Many of these are compounds that tend to inhibit the action of a very wide range of proteins. Others are fluorescent compounds, which appear as false positives in some of the standard assay techniques. They are typically fairly small molecules. These compounds will often show up as being positive for activity in initial assays of diverse libraries of compounds. However, they are generally not very useful as the basis of a drug design effort, owing to their promiscuous nature. Binding to a large variety of proteins means that there would be a wide range of side effects, probably severe, associated with utilizing them as drugs. A compilation of compounds that have been reported as promiscuous binders is presented in Fig. 2.4.

Sometimes, researchers use promiscuous binders to jump-start drug design efforts. This is usually done with the intent of using these compounds to map out a pharmacophore, then switching to development of other chemical series that could satisfy the same pharmacophore. It may also be possible to add

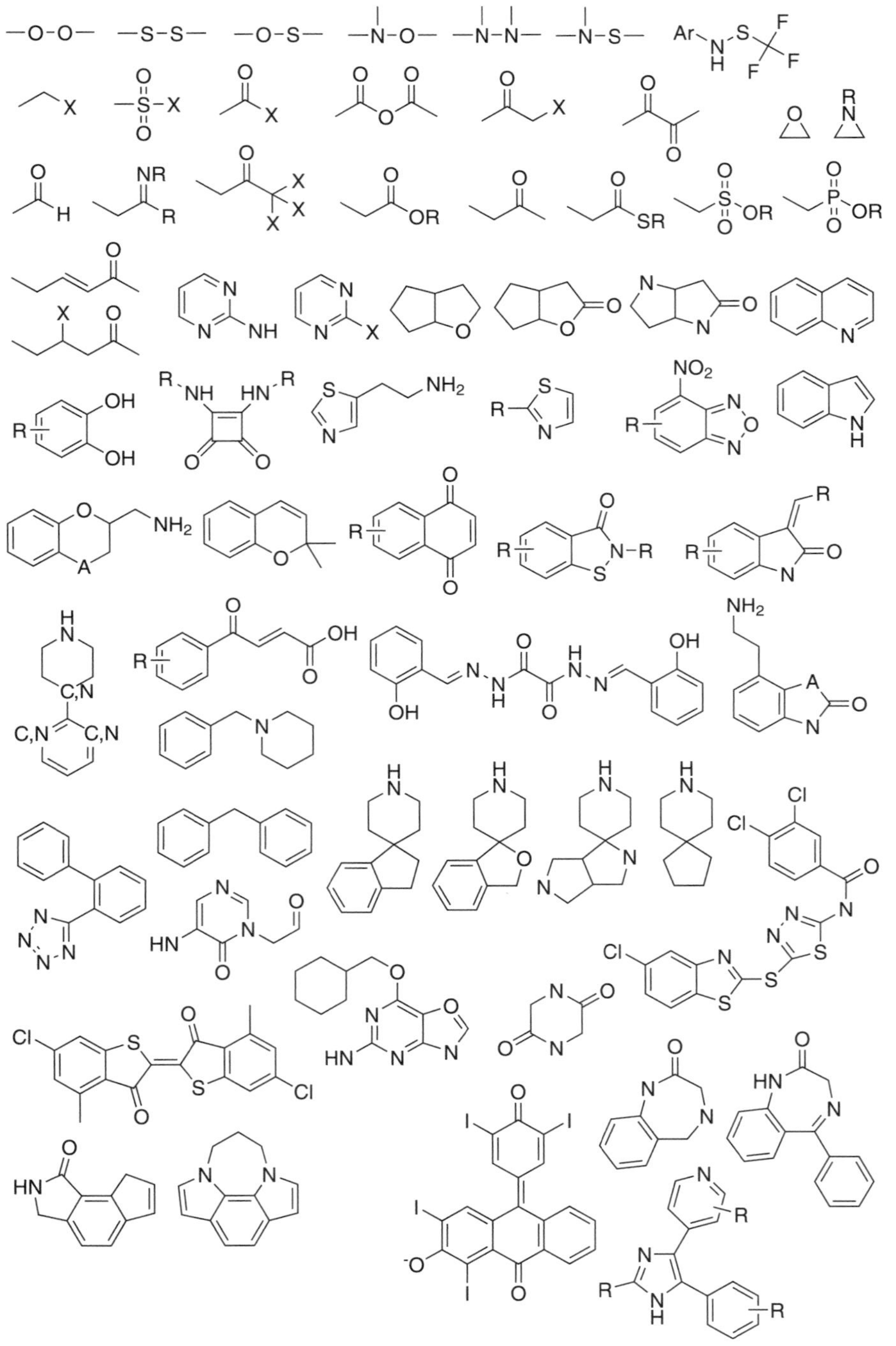

**Figure 2.4** Promiscuous compounds are compounds that inhibit a variety of unrelated protein targets.

enough additional functional groups to get the desired specificity—but not always. The term "privileged structures" is often used when using promiscuous compounds for this purpose.

There are also promiscuous proteins that bind a wide range of compounds. For example, AcrB is a protein that removes toxins from bacteria, such as *Escherichia coli*. This protein binds a wide range of molecules that should not be present in the bacterium. This is a mechanism for protecting the bacterium from antibiotics. Cytochrome P450 enzymes also bind a range of molecules for the purpose of detoxifying the blood. Promiscuous binding proteins are generally avoided as drug targets. If chosen as targets, they pose special problems for drug designers.

Using peptides and other naturally occurring compounds as drugs can lead to bioavailability problems. The human body is designed to breakdown foreign proteins in order to utilize them as a food source. Thus, using a peptide as a drug is often ineffective, since the body will break it down before it can have the desired drug effect. Therefore, many drugs have sizes, shapes, and hydrogen bond donors and acceptors that make them look very much like peptides, even though they are not formed from naturally occurring amino acids.

Many drugs have chiral centers. In almost all cases, the enantiomers do not interact with the body in the same way. Drug designers cannot avoid working with enantiomers, as chirality may be needed to obtain active compounds. However, designers should be aware that having a large number of chiral centers could greatly increase manufacturing costs.

In the past, both enantiomers would often be tested and approved for use as a racemic mixture. In recent years, the FDA has made a push to prevent racemic mixtures from being approved as pharmaceutical treatments. This is in response to a number of cases that have arisen over the years. The most widely published of these is the thalidomide tragedy.

Thalidomide was a drug sold to alleviate symptoms of morning sickness in pregnant women in the late 1950s and early 1960s. It was sold in Europe and many other countries around the world, but not the United States. Thalidomide is a chiral compound that was sold as a racemic mixture. One enantiomer alleviates morning sickness. The other enantiomer can cause severe birth defects, particularly if taken in the first trimester of pregnancy. Even when a single enantiomer is administered, this particular compound can undergo conversion to the other enantiomer in the body. Approximately 12,000 babies were born with severe deformities, such as missing limbs, and hands and feet with a flipper-like appearance. About 8000 of those born with these deformities survived past the first year of life. Thalidomide was never approved for use in the United States owing to the FDA's approval process being more stringent than that in other countries, and because of to the vigilance of the FDA reviewer assigned to the case, Frances Oldham Kelsey.

### 2.2.2 Bioavailability and Toxicity

Drug designers must carry out a delicate balancing act between efficacy and bioavailability. Ideally, drugs should be orally bioavailable so that they can be administered in tablet form. If a compound is too polar, it will not enter the bloodstream through passive intestinal adsorption, the primary mechanism for oral bioavailability. If the compound is too lipophilic, it will be eliminated from the bloodstream by the liver too quickly. Thus, the rule of thumb is:

> *A drug should be just lipophilic enough to reach the target, and no more lipophilic than necessary.*

There are, of course, time-released tablets, prodrug formulations, and other delivery mechanisms for alleviating bioavailability issues. However, drug designers will be urged to optimize bioavailability as much as is practical without sacrificing too much efficacy. If the drug has acceptable bioavailability as it is, then the manufacturing process will generally be less complex and costly.

The liver may eliminate a compound from the body as it is. Or it may break down the compound, usually through the action of various cytochrome P450 (CYP) enzymes. Some drug development firms are starting to incorporate a host of CYP interactions with the computational drug design process, to identify problems with excessively rapid elimination early in the development process. For more discussion of bioavailability, see Chapter 19.

Because CYP enzymes are responsible for the metabolism of many drugs, drug interactions can occur when one of the drugs affects the activity of these enzymes. This is a consideration when a patient is taking multiple drugs. Taking drugs that are CYP inhibitors generally causes the concentrations of other drugs in the bloodstream to increase, thus decreasing the needed dosage. Taking drugs that are CYP inducers tends to cause the patient to require larger doses of other drugs.

Because the liver functions to eliminate a wide range of compounds from the bloodstream, many drugs interact with the liver adversely, leading to liver toxicity (hepatotoxicity). The very molecular properties that help a compound stay in the bloodstream without being broken down by the liver can cause the compounds to be suicide inhibitors that harm the enzymes in the liver. For compounds that are given for a short period of time, such as antibiotics or anesthetics, a moderate amount of hepatotoxicity is acceptable. For drugs taken frequently over the patient's life, such as painkillers or cold medications, hepatotoxicity becomes a major concern in drug development, as it will be a significant factor in FDA approval of the drug.

Other types of toxicity are less of a concern in drug design—although that does not mean that they can be completely ignored. For example, very few drugs exhibit any neurotoxicity. This is because most drugs are moderately high molecular weight compounds that mimic small amino acid sequences. Such compounds are too large to fit in the sodium channel, which is the target for most neurotoxins. Most potential drugs are, however, tested for carcinogenicity and mutagenicity.

All substances are poisonous if taking in a sufficiently high dosage. They only differ in the dosage necessary to see a toxic effect. Toxicity is usually quantified by the lethal dose value $LD_{50}$, which is the amount necessary to kill 50% of test subjects. The units of $LD_{50}$ are usually mg/kg, meaning the number of milligrams of compound per kilogram of the subject's weight. For drug design purposes, the toxicities most frequently tested in animals are acute (single dosage) and chronic (multiple dosage), given both orally and by injection into the bloodstream. During the drug design process, determination of cytotoxicity is sometimes performed as a convenient cell-based assay. Table 2.1 lists some acute oral toxicity values for comparison. Toxicity studies are carried out to determine the safety margin between the dosage necessary for efficacy and the toxic dosage.

Cells are particularly sensitive to mutagens during mitosis. Thus, many drugs that can be administered to healthy adults cannot be given to pregnant women or young children. Mutagenicity is therefore usually included in drug design and assay testing. The first step in mutagenicity testing is the Ames test, a cell-based assay for mutagenicity. The computational tests for mutagenicity are QSAR models designed to predict the results of the Ames test. A more detailed discussion of toxicity and mutagenicity prediction is in Chapter 19.

**TABLE 2.1 Acute Oral Toxicity**

| Compound | $LD_{50}$ (mg/kg) |
|---|---|
| Botulinum toxin A | 0.0014 |
| Parathion | 8.3 |
| Strychnine | 30 |
| Mercury | 37 |
| Nicotine | 55 |
| DDT | 113 |
| Gasoline | 150 |
| Warfarin | 186 |
| Caffeine | 200 |
| Chlordane | 382 |
| Aspirin | 1200 |
| Malathion | 1375 |
| Table salt | 3300 |
| Ethanol | 4500 |

### 2.2.3 Drug Side Effects

Nearly all drugs exhibit some type of undesirable side effect. There are many types of potential side effects, such as addiction, drowsiness, allergic reactions, and dry mouth. In some cases, side effects can be fatal. In many cases, drugs are approved when side effects occur in only a very small percentage of the population, or are relatively mild. However, somewhat severe side effects can be tolerated if the drug is extending life expectancy for patients with an otherwise fatal condition.

Unfortunately, side effects are usually not identified until human clinical trials. This means that a very large sum of money has been spent already when these problems are identified. The cost becomes even higher if the drug fails clinical trials. The pharmaceutical industry would gladly embrace a technology that could identify side effects of a drug before it entered human clinical trials, but at present no reliable solution to this problem exists.

One option for early identification of side effects that is utilized is the observation of animals in efficacy studies. Unfortunately, this generally only allows identification of side effects that cause behavioral changes in the animals. For example, drugs causing moderate drowsiness can be identified by documenting how much time the animals spend asleep. Many side effects cannot be identified or may not be present in animals. Likewise, there may be side effects in animals that are not exhibited in humans.

In theory, it should be possible to predict side effects by computationally identifying other targets in the body with which the drug could interact. These targets could be identified through the use of bioinformatics, proteomics, or high speed docking techniques. Once a side-effect target is identified, the metabolic pathways in which the target participates could indicate what type of side effects would be observed. Unfortunately, this schema works poorly, if at all, in practice, owing to the current limited knowledge of the proteome, metabolic pathways, and expression profiles.

What is routinely done is to run docking studies and assays against the proteins that show the highest similarity to the target. The assumption is that if the drug has sufficient specificity to bind to the intended target, and not to the protein most similar to that target, then there should be a reasonable expectation that side effects should be minimized. This is a good first approximation, but not rigorously true, as can be seen from the example above of binding to cytochrome P450 enzymes, which promiscuously bind a large selection of substrates.

### 2.2.4 Multiple Drug Interactions

Some drugs will have interactions with other drugs, thus preventing both from being taken simultaneously, or forcing physicians to adjust the dosage accordingly. Cross reactions are seldom considered in the research and development phase unless the company plans to market a drug with fewer cross reactions than their competitor's product. There is no simple test for side effects of administering multiple drugs to a patient. Most of this data is gained from human clinical trials. Some initial indications can be suggested by looking at drug interaction with cytochrome P450 enzymes. Beyond that, there are only a few vague thumb rules to guide clinical testing, such as "*do not mix depressants*" and "*everything interacts with blood thinners.*"

Some drug interactions are particularly difficult to anticipate. This is illustrated rather dramatically by the Fen-Phen story. Fen-Phen was a diet drug that consisted of a mixture of two existing obesity drugs fenfluramine (trade name Pondimin) and phentermine. The different drugs displayed opposing side effects. Fenfluramine is an antihistamine that makes patients drowsy and phentermine is a mild stimulant that increases the body's metabolism while suppressing appetite. The mixture was marketed together in an attempt to give an obesity drug with minimal side effects. The combination of the drugs also gave more dramatic weight loss than either drug taken separately.

Unfortunately, the mixture of drugs in Fen-Phen gave problems that had not been identified from the use of the drugs individually. The most severe of these side effects were the occurrence of serious cardiac valvular disease, and primary pulmonary hypertension. In September 1997, at the FDA's request, Fen-Phen and fenfluramine were removed from the market, as was dexfenfluramine (trade name Redux), which is one of the enantiomers of which fenfluramine is the racemic mixture. After that, a number of law suits regarding long-term affects of Fen-Phen-induced heart disease were filed against American Home Products, the parent company that marketed Fen-Phen. American Home Products subsequently agreed to a class action settlement valued at $4.75 billion.

## 2.3 METRICS FOR DRUG-LIKENESS

Up to this point, this chapter has discussed the concerns that must be addressed in designing a drug in general terms. This section will focus on molecular structure. A number of metrics have been developed for identifying whether compounds are drug-like. These criteria translate directly or indirectly into desired molecular size, functional groups, etc. Note that drug-likeness metrics do not indicate that a compound will be a good drug for any disease. However, compounds that do not meet drug-likeness criteria often fail to be good drugs due to poor bioavailability, excessive toxicity or other concerns.

The terms "drug-likeness" and "lead-likeness" have both appeared in the recent literature. These terms are used in conflicting and overlapping ways from one article to another. Some authors distinguish "drug-likeness" as describing structural features, such as certain types of functional groups, while they use the term "lead-likeness" when describing physical properties, such as polar surface area. Other authors use the term "lead-likeness" to mean the use of more restrictive criteria than those used for "drug-likeness," although they may be looking at the same features of molecular structure. Some authors treat the two as being synonymous. This text will use the term "drug-likeness" for all of the above types of molecular analysis.

One of the criteria most frequently used for drug-likeness is the Lipinski rule of fives. This rule was not created to be a drug-likeness criterion, but rather as a criterion for identifying molecules that will be orally bioavailable. However, pharmaceutical companies push for orally bioavailable drugs wherever possible, because patients, doctors, and insurance companies prefer drugs in tablet form. Thus, the two criteria are synonymous for many drug design projects. The Lipinski rule of fives states that for a drug to be orally bioavailable, it should meet at least three of the following criteria:

- The compound should have **5** or fewer hydrogen-bond donors.
- The compound should have a molecular weight of **5**00 or less.
- The substance should have a calculated log $P$ (Clog$P$) less than or equal to **5** (or Mlog$P \leq 4.15$).
- The compound should have 10 or fewer hydrogen-bond acceptors.

Lipinski et al., also note that compound classes that are substrates for biological transporters are exceptions to the rule. Thus, these criteria only predict passive intestinal absorption, not active transport. Other oral bioavailability metrics can also be utilized as drug-likeness metrics. For more information on oral bioavailability prediction, see Chapter 19.

Many drug likeness metrics have included a log $P$ criterion. Programs to compute log $P$ quickly and reasonably accurately have been available for many years. A number of researchers have noted that it would be more appropriate to use log $D$, which takes into account the ionization states of the molecule at a specified pH. Some use $\log D_{5.5} \leq 5$ or $\log P < 6.25$ as a substitute term in the Lipinski rules. There are now programs for quickly and reliably computing log $D$, such as that from ACD/Labs.

Some researchers will use similar rules for identifying reasons why a given compound might not make a viable drug. For example, Schneider and Baringhaus noted the following:

- A drug-like molecule should not have more than 5 rotatable bonds (sometimes referred to as the fifth Lipinski rule).
- A drug-like molecule should not possess a polar surface area exceeding 120 Å$^2$, since it would then not be sufficiently lipophilic to enter the bloodstream through passive intestinal absorption in the small intestine.

- A drug-like molecule should not have an aqueous solubility (log $S$) less than $-4$.

It should be noted that a perusal of the structures of approved drugs will reveal a noticeable range of molecular weights, topologies, and functional groups. In general, drugs that treat central nervous system disorders tend to be at the lower end of the molecular weight range, with proportionally fewer heteroatoms and few, if any, very polar functional groups. This may be in part due to the additional constraint of having to pass through the blood–brain barrier. Other orally administered low molecular weight drugs, such as some of the broad spectrum antibiotics, tend to have a greater number of heteroatoms and thus more polarized charge distributions. At the higher end of the molecular weight range are steroids, hormones, and macrocyclic antibiotics.

A set of rules similar to the Lipinski rules was developed by Oprea, who analyzed a somewhat larger database of molecules. He found that 70% of drugs contain the following.

- 0–2 hydrogen-bond donors
- 2–9 hydrogen-bond acceptors
- 2–8 rotatable bonds
- 1–4 rings

Of course, it would be unwise to use these criteria so stringently that the other 30% of profitable drugs on the market were excluded from consideration. However, it is reasonable to assume that molecules become less drug-like (for most applications) as they display larger deviations from these metrics.

Oprea, Bologa, and Olah performed an analysis of a large number of approved drugs from MDL's Drug Data Report and the Physicians Desk Reference to develop the following lead-likeness metric (note the similarity to the Lipinski rules, except with tighter boundaries):

- molecular weight $\leq 460$
- $4 \leq \text{ClogP} \leq 4.2$
- $\log S_w \leq -5$
- number of rotatable bonds $\leq 10$
- number of hydrogen bond donors $\leq 5$
- number of hydrogen bond acceptors $\leq 9$

Oprea et al., note that most drug-likeness and lead-likeness metrics are developed by comparing databases of drugs with databases of commercially available chemicals. Since approved drugs are typically more complex than other chemicals on the market, these metrics become in some part a measure of structural complexity. They also point out that some other properties to look for are log $D_{7.4}$ (log $P$ for ionization states at pH 7.4) between 0 and 3, little or no binding to cytochrome P450 enzymes, and plasma protein binding below 99.5%.

Another approach to giving researchers an understanding of drug-likeness is to look at structural motifs found in existing drugs. This is not necessarily very useful as a library screening tool, but it can be very relevant to determining what types of functional groups should or should not be included in combinatorial library design and other synthesis efforts.

Several papers by Bemis and Murcko present a topological shape analysis and structural analysis of the approved drugs in the Comprehensive Medicinal Chemistry (CMC) database of approved drugs, excluding some nondrug listings such as imaging agents and dental resins. The topological analysis looked at connectivity only, ignoring element and bond order. Of the drugs analyzed, 94% contained rings, and the remaining 6% were acyclic molecules. There were 1179 different ring topologies, 783 of which appeared in only a single drug. There were 32 ring/linker topologies that appeared in 20 compounds or more. Two of these contained a single ring, with the largest topological category being drugs with a single six-membered ring. Of the 32 most common topological categories, 18 contain fused ring systems, only 1 of which has a bridgehead topology. Of the 32 categories, 18 contain two ring systems linked together.

The subsequent analysis of drug frameworks by Bemis and Murcko takes into account the identity of heteroatoms and bond orders. Forty-one such frameworks are represented at least 10 times in the database. Of these, 25 have noncarbon atoms in the rings. These studies certainly illustrate the importance of rings and heterocyclic compounds in drug design. They also suggest some specific ring systems that drug designers would be wise to understand and consider.

A subsequent analysis of side chains by Bemis and Murcko identified 1246 unique side chains, with an average of four side chains per molecule. The 20 most commonly occurring side chains were found 11,000 times in the dataset. Of the 46 most commonly occurring side chains, 37 have heteroatoms. Figure 2.5 shows a compilation of functional groups most frequently found in many drugs, per the sources listed in this chapter.

Some companies also use a "black list" of structural motifs that they want to avoid incorporating in drugs (see the book by Schneider and Baringhaus in the Bibliography). Some groups that are blacklisted by some companies are thioureas, esters, amides, disulfides, thiols, β-lactams, *O*-nitro, alkoxypyridinium, benzophenone, oxadiazine, fluorenone, and acylhydroquinones. Note that some of these, such as thiols and β-lactams are found in a few

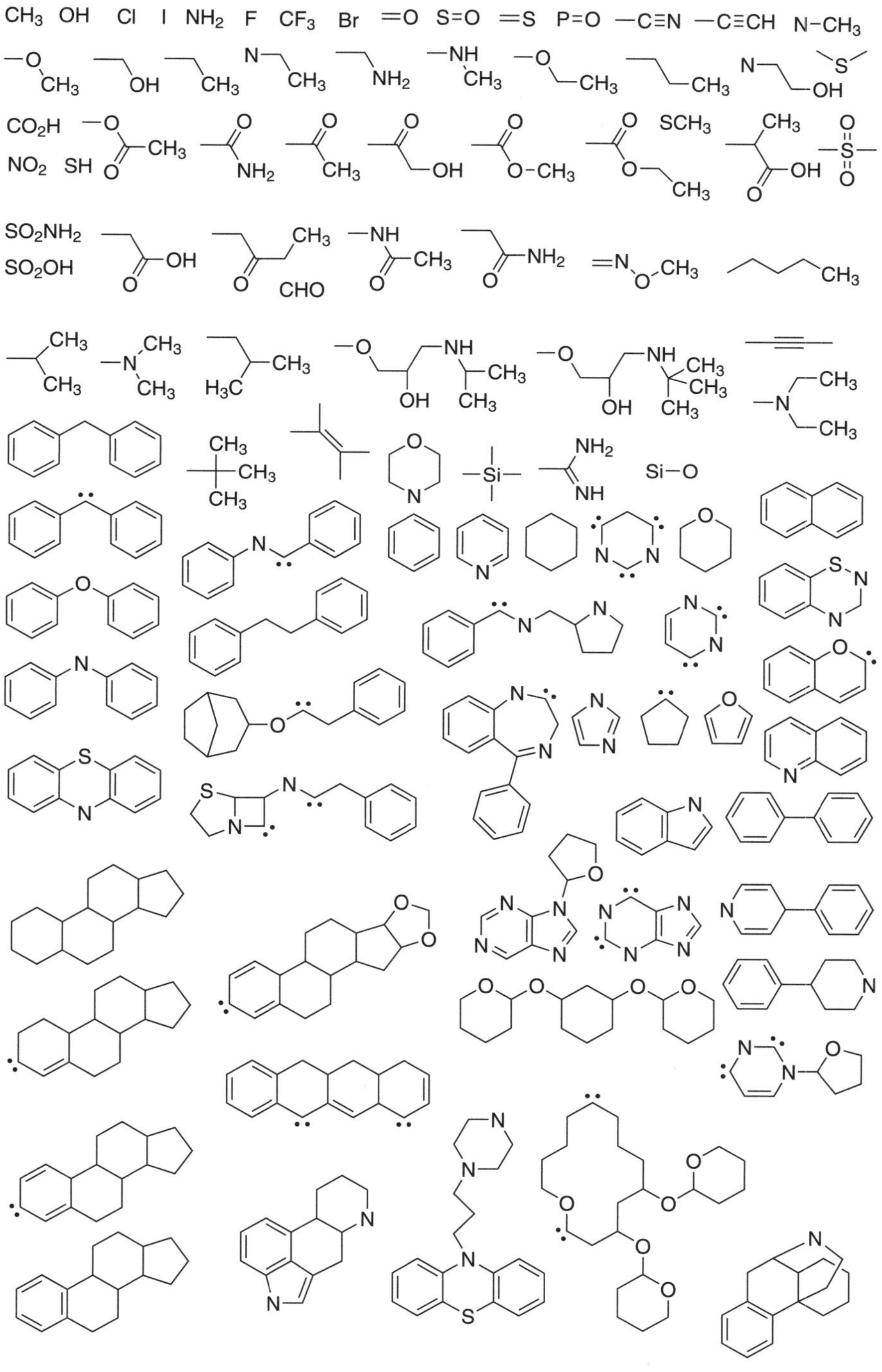

**Figure 2.5** Structural motifs most frequently found in drugs.

commercially available drugs. Some of these discrepancies can be due to scientific concerns, such as the wide range of bacteria exhibiting resistance to β-lactam antibiotics. Esters and amides are readily broken down in the body by esterases and proteases, thus making them poor choices for a drug, but good options for forming a prodrug. There can also be differences simply because of the classes of products under development at a particular pharmaceutical company.

The property distributions above are easily understood descriptions of what properties are typical in drugs. These distributions are used in drug design work, although the user must be wary of taking them too literally, as illustrated by Oprea's approach, which works for only 70% of known drugs. Methods exist for much more accurately dividing large compound libraries into drug-like and nondrug-like groups. These techniques tend to be more mathematical, and less easily understood. They include techniques based on molecular fingerprints, tree-based classification, neural networks, expert systems, knowledge-based (heuristic) algorithms, support vector machines, and other techniques. One technique, due to Takaoka et al., was parameterized by asking chemists to rate how similar various compounds are to one another. These techniques tend to be based on a philosophy of finding something that works, even if it may be unclear what the various parameters in the model are saying about the chemistry.

Another class of drug-likeness methods are actually multiple models, each with a physical interpretation, taken together to give an overall indication of drug-likeness. For example, a researcher may use a passive intestinal absorption model, a metabolic stability model, and a toxicity model, in order to find compounds that are acceptable under all three criteria. This makes the results easier to interpret, and allows researchers to see why a particular compound failed. However, this does not necessarily make the final prediction any more accurate. These tools are often classified as pharmacokinetic modeling tools, even though in practice they may be utilized as drug-likeness criteria. For more information on these techniques, see Chapter 19.

In practice, drug designers tend to use multiple tools at different points in the design process. When performing initial screenings, researchers will use techniques that err on the side of including many inactive compounds in order to avoid leaving out any that might be useful. The opposite extreme is seen when preparing to go into animal trials. At that point, researchers may start with a large list of active compounds and use very restrictive criteria to narrow it down to a very small list of compounds that are the best of the best.

The usefulness of drug-likeness metrics often depends upon the way in which they were developed, and the way in which they are to be used. Many of the single-number drug likeness metrics are developed by generating a database of drugs and nondrugs and trying to find metrics to separate the two. The nondrugs are usually taken from listings of chemicals available for sale. The compounds available for sale are typically those ideal for use in synthesis. As such, commercial chemicals tend to be smaller molecules than most drugs, and have one or two functional groups instead of the multiple functional groups on most drugs, and many commercial compounds are too hydrophobic to make good drugs. However, such metrics may not be able to distinguish between known drugs and natural products that are neurotoxins (or other substances that could enter the body as a drug does).

In theory, the methods that have several physical models, such as oral bioavailability and toxicity, should be able to avoid such problems. However, these methods are prone to other shortcomings. The most notable problem is a tendency to label many known, profitable drugs as being not drug-like.

## 2.4 EXCEPTIONS TO THE RULES

Much of this chapter has been devoted to discussing what properties make a compound a good drug. However, there are numerous examples of successful drugs that do not obey all of these "rules." The following paragraphs discuss some of these exceptions. Noting why these examples were acceptable drugs gives some insight into situations in which researchers may reasonably be able to go outside the conventional bounds of drug design.

Cisplatin (Fig. 2.6) is a cancer chemotherapy drug. With only 11 atoms, it is smaller and less complex than most drugs. As implied by the discussions of specificity earlier in this chapter, one might expect this to result in unwanted side effects. Indeed, cisplatin use often causes severe nausea, hair loss, weight loss (perhaps indirectly from the nausea), kidney damage, hearing loss, sometimes some nerve damage, and electrolyte disturbances. However, cisplatin was originally the only hope to extend the life of patients with many forms of cancer. Thus, the severe side effects were acceptable. Today, cisplatin is still used for some forms of cancer, often in small doses mixed with other drugs having less severe side effects.

Some treatments are actually proteins, which are much larger than the typical drug. For example, Gaucher's disease is a genetic mutation in which the patient's body does not produce the enzyme glucocerebrosidase. The

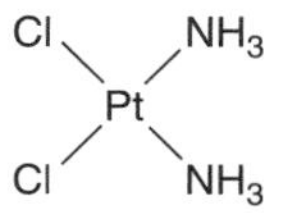

Cl NH$_3$
Pt
Cl NH$_3$

**Figure 2.6** The cancer chemotherapy drug cisplatin.

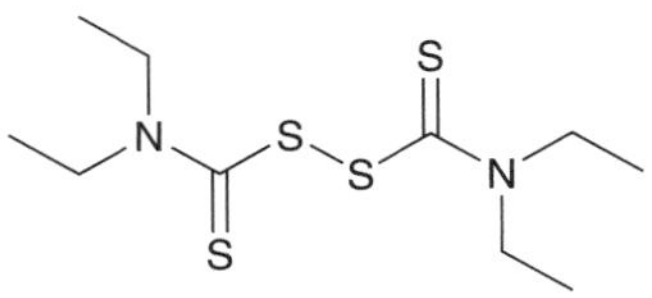

**Figure 2.7** Disulfiram. This drug is effective for treating alcoholism because of the severe reaction that occurs when it is taken with alcohol.

preferred treatment for Gaucher's disease is enzyme replacement therapy. Another medical application of proteins is the injection of botulinum toxin type A (trade name Botox) for cosmetic applications, severe underarm sweating, cervical dystonia, uncontrollable blinking, and misaligned eyes. The toxin is a 900 kDa protein consisting of two protein strands linked by a disulfide bridge. In all of these applications, the injection partially paralyzes the tissue. The large size of the protein prevents it from migrating throughout the body and causing unwanted effects elsewhere.

Another interesting drug is disulfiram (Fig. 2.7), also called tetraethylthiuram disulfide (trade name Antabuse). This compound has an unusual dithiocarbamate functional group. One would generally be concerned about the potential for toxicity or severe side effects from such a chemical structure. However, disulfiram is used as a medication specifically because of its toxic effects. Disulfiram is administered to patients as part of the treatment for alcoholism. Disulfiram blocks the oxidation of alcohol at the acetaldehyde stage, thus causing the patient to become violently ill if even slight amounts of alcohol are ingested while disulfiram is in the bloodstream. Several species of mushroom contain very similar compounds, which also show a severe cross-reaction with alcohol.

Another example of a useful side effect comes from the Viagra development story. Viagra (trade name for sildenafil) was originally developed as a drug for hypertension. When the researchers discovered that the drug target was prevalent in vascular muscle walls, the project shifted the goal from a hypertension treatment to an angina treatment. When the drug entered human clinical trials, it became clear almost immediately that it was less effective than the existing nitroglycerin treatments. In the course of those initial clinical

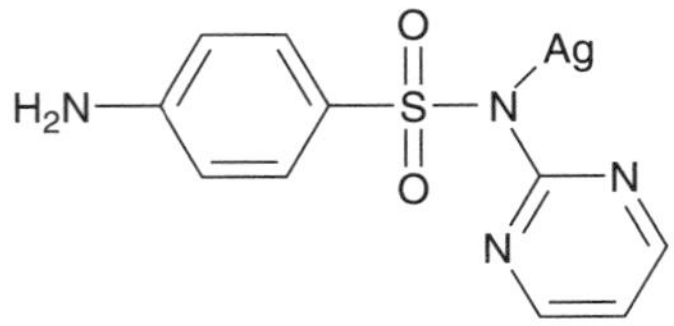

**Figure 2.8** Silver sulfadiazene. This drug is for topical use only.

trials, the hospital staff noted that one side effect was an improvement in erectile dysfunction. This observation was taken seriously, and eventually lead to Viagra, which is now a blockbuster drug, meaning that sales have grossed over a billion dollars.

Normally, one would not consider a silver compound such as silver sulfadiazine (Fig. 2.8) as a drug candidate, owing to the potential for heavy metal poisoning. Indeed, the toxicity of silver sulfadiazine is the key to its usefulness as an antimicrobial. However, this compound is administered only topically as Silvadene Cream 1%. Because of the topical application, very little of the compound enters the body. This makes it useful for the treatment of wound sepsis in patients with second- and third-degree burns. This illustrates that researchers should keep the application in mind when determining what molecular properties can be considered drug-like.

These examples serve to show that there are a variety of exceptions to the typical requirements of drug design. Those requirements are a rich set of varied, and sometimes conflicting, needs, including activity, bioavailability, toxicity, specificity, and sometimes even cost. This is responsible for the true complexity of drug design.

## BIBLIOGRAPHY

### Assay Information

Minor LK, editor. Handbook of Assay Development in Drug Discovery. Boca Raton, FL: CRC Press; 2006.

Vogel H, editor. Drug Discovery and Evaluation: Pharmacological Assays. Heidelberg: Springer; 2007.

### Animal Testing Information

Altweb: The Global Clearinghouse for Information on Alternatives to Animal Testing. Available at http://altweb.jhsph.edu. Accessed 2006 Apr 20.

Transgenic Animal Web: Links to Useful Resources. Available at http://www.med.umich.edu/tamc/links.html. Accessed 2006 Apr 20.

### The Clinical Testing Process

Eckstein J. ISOA/ARF Drug Development Tutorial. Available at http://www.alzforum.org/drg/tut/ISOATutorial.pdf. Accessed 2006 Apr 27.

US Food and Drug Administration Center for Drug Evaluation and Research. Guidance Documents. Available at http://www.fda.gov/CDER/GUIDANCE. Accessed 2006 Apr 27.

## Drug Binding

Ehrlich P. Über den jetzigen Stand der Chemotherapie. Ber Dtsch Chem Ges 1909; 42: 17–47.

Fischer E. Einfluss der Configuration auf die Wirkung der Enzyme. Chem Ber 1894; 27: 2985–3189.

Fischer E, Thierfelder H. Verhalten der verschiedenen Zucker gegen reine Hefen. Chem Ber 1894; 27: 2031–2037.

Koshland DE Jr. Application of a theory of enzyme specificity to protein synthesis. Proc Natl Acad Sci USA 1958; 44: 98–104.

Stinson SC. Counting on chirality. Chem Eng News 2000; 76(38): 83–104.

## Promiscuous Binding Compounds

Bondensgaard K, Ankersen M, Thogersen H, Hansen BS, Wulff BS, Bywater RP. Recognition of privileged structures by G-protein coupled receptors. J Med Chem 2004; 47: 888–899.

Crisman T, Parker C, Jenkins J, Scheiber J, Thoma M, Kang Z, Kim R, Bender A, Nettles J, Davies J, Glick M. Understanding false positives in reporter gene assays: In silico chemogenomics approaches to prioritize cell-based HTS data. J Chem Inf Model 2007; 47: 1319–1327.

DeSimone RW, Currie KS, Mitchell SA, Darrow JW, Pippin DA. Privileged structures: Applications in drug discovery. Comb Chem High Throughput Screen 2004; 7: 473–493.

Goodnow RA, Gillespie P, Bliecher K. Chemoinformatic tools for library design and the hit-to-lead process: A user's perspective. In: Oprea TI, editor. Chemoinformatics in Drug Discovery. Weinheim, Wiley-VCH; 2004. p. 381–435.

Guo T, Hobbs DW. Privileged structure-based combinatorial libraries targeting G protein-coupled receptors (GPCRs). Assay Drug Dev Technol 2003; 1: 579–592.

Harris W, Jones P, Taniguchi M, Tsujii S, von der Saal W, Zimmermann G, Schneider G. Development of a virtual screening method for identification of "Frequent Hitters" in compound libraries. J Med Chem 2002; 45: 137–142.

Horton DA, Bourne GT, Smythe ML. Exploring privileged structures: The combinatorial synthesis of cyclic peptides. J Comput Aided Mol Des 2002; 16: 415–430.

Horton DA, Bourne GT, Smythe ML. The combinatorial synthesis of bicyclic privileged structures or privileged substructures. Chem Rev 2003; 103: 893–930.

Mason JS, Morize I, Menard PR, Cheney DL, Hulme C, Labaudiniere RF. New 4-point pharmacophore method for molecular similarity and diversity applications: Overview of the method and applications, including a novel approach to the design of combinatorial libraries containing privileged substructures. J Med Chem 1999; 42: 3251–3264.

McGovern SL, Saselli E, Grigorieff N, Shoichet BK. A common mechanism underlying promiscuous inhibitors from virtual and high throughput screening. J Med Chem 2002; 45: 1712–1722.

McGovern SL, Shoichet BK. Kinase inhibitors: Not just for kinases anymore. J Med Chem 2003; 46(8): 1478–1483.

Müller G. Medicinal chemistry of target family-directed masterkeys. Drug Discov Today 2003; 8: 681–691.

Nicolaou KC, Pfefferkorn JA, Roecker AJ, Cao GQ, Barluenga S, Mitchell HJ. Natural product-like combinatorial libraries based on privileged structures. 1. General principles and solid-phase synthesis of benzopyrans. J Am Chem Soc 2000; 122: 9939–9953.

Patchett AA, Nargund RP. Privileged structures—an update. Annu Rep Med Chem 2000; 35: 289–298.

Rishton GM. Nonleadlikeness and leadlikeness in biochemical screening. Drug Discov Today 2003; 8: 86–96.

Rishton GM. Reactive compounds and in vitro false positives in HTS. Drug Discov Today 1997; 2: 382–384.

Roche O, Schneider P, Zuegge J, Bleicher K, Danel F, Gutknecht E, Rogers-Evans M, Neidhard W, Stalder H, Dillon M, Sjoegren E, Fotouhi N, Gillespie P, Goodnow R, Savchuk NP, Tkachenko SE, Balakin KV. Rational design of GPCR-specific combinatorial libraries based on the concept of privileged substructures. In: Oprea TI, editor. Chemoinformatics in Drug Discovery. Weinheim, Wiley-VCH; 2004. p. 287–313.

## Drug Likeness

Ajay. Predicting drug-likeness: Why and how? Curr Top Med Chem 2002; 2: 1273–1286.

Bemis GW, Murcko MA. The properties of known drugs. 1. Molecular frameworks. J Med Chem 1996; 39: 2887–2893.

Bemis GW, Murcko MA. The properties of known drugs. 2. Side chains. J Med Chem 1999; 42: 5095–5099.

Biswas D, Roy S, Sen S. A simple approach for indexing the oral druglikeness of a compound: Discriminating druglike compounds from nondruglike ones. J Chem Inf Model 2006; 46: 1394–1401.

Ertl P. Cheminformatics analysis of organic substituents: Identification of the most common substituents, calculation of substituent properties, and automatic identification of drug-like bioisosteric groups. J Chem Inf Comput Sci 2003; 43: 374–380.

Goodnow RA, Gillespie P, Bliecher K. Chemoinformatic tools for library design and the hit-to-lead process: A user's perspective. In: Oprea TI, ed. Chemoinformatics in Drug Discovery. Weinheim: Wiley-VCH; 2004: 381–435.

Lipinski CA, Lombardo F, Dominy BW, Feeney PJ. Experimental and computational approaches to estimate solubility and permeability in drug discovery and development settings. Adv Drug Deliv Rev 1997; 23: 3–25.

Mannhold R, Kubinyi H, Folkers G, editors. Molecular Drug Properties: Measurement and Prediction (Methods and Principles in Medicinal Chemistry). Weinheim: Wiley-VCH; 2007.

Müller KR, Rätsch G, Sonnenburg S, Mika S, Grimm M, Heinrich N. Classifying "drug-likeness" with kernel-based learning methods. J Chem Inf Model 2005; 45: 249–253.

Nogrady T, Weaver DF. Medicinal Chemistry: A Molecular and Biochemical Approach. 3rd ed. Oxford: Oxford University Press; 2005.

Oprea TI, Bologa C, Olah M. Compound selection for virtual screening. In: Alvarez J, Shoichet B, editors. Virtual Screening in Drug Discovery. Boca Raton, FL: CRC Press; 2005. p. 89–106.

Oprea TI. Property distribution of drug-related chemical databases. J. Comput Aided Mol Des 2000; 14: 251–264.

Schneider G, Baringhaus KH. Molecular Design. Weinheim: Wiley-VCH; 2008.

Takaoka Y, Endo Y, Yamanobe S, Kakinuma H, Okubo T, Shimazaki Y, Ota T, Sumiya S, Yoshikawa K. Development of a method for evaluating drug-likeness and ease of synthesis using a dataset in which compounds are assigned scores based on chemists' intution. J Chem Inf Comput Sci 2003; 43: 1269–1275.

Walters WP, Murcko MA. Prediction of "drug-likeness." Adv Drug Deliv Rev 2002; 54: 255–271.

### Toxicity

Casarett LJ, Amdur MO, Klaassen CD, Doull J, editors. Casarett and Doull's Toxicology: The Basic Science of Poisons. 5th ed. Columbus, OH: McGraw-Hill; 1996.

### Drug Side Effects and Interactions

Thalidomide. Center for the Evaluation of Risks to Human Reproduction (CERHR), Available at http://cerhr.niehs.nih.gov/common/thalidomide.html. Accessed 2008 Nov 16.

Kling J. From hypertension to angina to Viagra. Mod Drug Discov 1998; 1: 31–38.

Fackelmann K. Diet drug debacle. Science News Online. Available at http://www.sciencenews.org/sn_arc97/10_18_97/bob2.htm. Accessed 2008 Nov 16.

### Protein Classification

Koehl P. Protein structure classification. Rev Comput Chem 2006; 22: 1–55.

### Examples of Exceptions to the Rules

Chabner BA, Amrein PC, Druker B, Michaelson MD, Mitsiades CS, Goss PE, Ryan DP, Ramachandra S, Richardson PG, Supko JG, Wilson WH. Antineoplastic agents. In: Brunton LL, Lazo JS, Parker KL, editors. Goodman & Gilman's Pharmacological Basis of Therapeutics. 11th ed. New York: McGraw-Hill: 2005. p. 1315–1404.

Duplay D. et al. Physicians' Desk Reference, 59th ed. Montvale: Thomson PDR; 2005. p. 2442 and 1792.

Internet Mental Health. Disulfiram. Available at http://www.mentalhealth.com/drug/p30-a02.html. Accessed 2008 Feb 14.

MedlinePlus. Botox. Available at http://www.nlm.nih.gov/medlineplus/botox.html. Accessed 2008 Feb 14.

NINDS Gaucher's Disease Information Page. Available at http://www.ninds.nih.gov/disorders/gauchers/gauchers.htm. Accessed 2008 Feb 14.

Rosenberg B, Van Camp L, Krigas T. Inhibition of cell division in *Escherichia coli* by electrolysis products from a platinum electrode. Nature 1965; 205: 698–699.

Additional references are contained on the accompanying CD.

# 3

# TARGET IDENTIFICATION

## 3.1 PRIMARY SEQUENCE AND METABOLIC PATHWAY

With the completion of the Human Genome Project, we now have the primary amino acid sequence for all of the potential proteins in a typical human body. However, knowledge of the primary sequence alone is not enough on which to base a drug design project. For example, the primary sequence does not tell when and where the protein is expressed, or how proteins act together to form a metabolic pathway. Even more complex is the issue of how different metabolic pathways are interconnected.

Ideally, the choice of which protein a drug will inhibit should be made based on an analysis of the metabolic pathways associated with the disorder that the drug is intended to treat. In reality, many metabolic pathways are only partially understood. Furthermore, intellectual property concerns may drive a company toward or away from certain targets.

Several databases of metabolic pathways are available. One is the MetaCyc database available at http://metacyc.org. Another is the KEGG Pathway Database available at http://www.genome.ad.jp/kegg/pathway.html, which is interconnected with the SEED Annotation/Analysis Tool at http://theseed.uchicago.edu/FIG/index.cgi. The ExPASy Biochemical Pathway page at http://theseed.uchicago.edu/FIG/index.cgi is a front end to the digital version of the Roche Applied Science "Biochemical Pathways" wall chart.

*Computational Drug Design*. By David C. Young

**Figure 3.1** Superimposed structures of the SARS protease protein obtained from crystallography (light/yellow), and a homology model (dark/blue). This is typical of a good homology model, in that the largest errors are in loops, while α-helices and β-sheets are reproduced most accurately. A color copy of this figure is contained on the accompanying CD.

These are far from providing a comprehensive listing of all the metabolic pathways in a given organism. However, they do serve as repositories for all of the information currently known, with the exception of some data that is proprietary to specific pharmaceutical companies.

A discussion of metabolic pathways is beyond the scope of this book. Therefore, the discussion in the rest of the book assumes that a target has been selected.

Structure-based drug design techniques are used when possible, since these have the highest success rate. Structure-based drug design requires knowledge not only of the target primary sequence, but also of the three-dimensional structure of the target. Therefore, the rest of this chapter is devoted to a discussion of techniques for determining the three-dimensional shape of proteins.

Of the techniques in the following discussion, crystallography is the preferred way of determining a target geometry, if possible. 2D NMR can give an accurate model for just the active site. Homology models can give either good or incorrect structures, depending upon the amount of similarity to proteins for which crystal structures are available. Protein folding is an absolute last resort, and researchers may choose to cancel a project rather than trust a protein folding model. Figure 3.1 gives an example of the structures from X-ray crystallography and a well-made homology model.

## 3.2 CRYSTALLOGRAPHY

Ideally, target protein structures should be determined by X-ray crystallography before starting a drug design project. Additional information about protein conformational changes can be obtained if structures both with and without a ligand soaked in the active site are obtained. These are referred to as a "soaked" structure and an "apo" structure. If both are not available, the soaked structure is preferred.

Unfortunately, proteins are notoriously difficult to crystallize, and membrane-bound proteins are far more difficult than other proteins. However, the information gained from crystallography is so valuable that extraordinary efforts and costs are justified. There are even "structural genomics" projects that are attempting to crystallize all known proteins.

Like any experimental measurement, protein structures are not exact. In fact, it is often the case that the measurement gives an electron density that shows a region where a given nonhydrogen atom is located. Often, hydrogen atoms cannot be located at all in the electron density of a protein crystal. This relative lack of resolution, compared with crystallography of small molecules, is due to the very small size of protein crystals. In order to obtain a usable structure, it is necessary to apply an optimization procedure to find atomic positions that simultaneously fit the electron density and minimize bond strain energy using a force field approach.

There are several measures of the quality of a crystal structure. The positional accuracy of each atom can be quantified by a *B*-factor, which is an indicator of the thermal motion of the atoms. It can also be an indication of heterogeneity in the crystal. A resolution value, in ångströms, is computed in order to give an indication of the overall quality of the structure. If the resolution is 1.0 Å, a plot of the electron density will clearly show the heavy atoms, for example making a phenyl group clearly recognizable. If the resolution is 2.0 Å, the shape of a functional group should be well resolved, but the individual atoms will have to be fit to that shape using a force field technique. If the resolution is 3.0 Å, the overall shape of the electron density for a functional group may be somewhat distorted. About 90% of protein crystal structures have a resolution greater than 1.6 Å. It should be noted that the quality of the structure, quantified as a resolution, is a global property in crystallography, and a local one in NMR.

Protein crystal structures will appear slightly different, depending upon the force field used to optimize them. Ideally, drug designers should utilize a structure optimized with the same force field that will be used in the design work, in order to obtain a consistent description with that force field. This is done by performing a simple downhill minimization of the crystal structure with the force field to be utilized. Note that optimizations *in vacuo* tend to give a structure that is slightly compressed owing to the lack of water molecules being included either explicitly or as a dielectric, so it is generally best to perform the optimization with a compound in the active site. This usually results in a structure that is the same conformer and indistinguishable to the naked eye unless the two structures are overlaid. If loops or side chains undergo significant changes in shape when minimized without the electron density constraints, this usually indicates that those sections were pushed into a less energetically stable position by the crystal packing forces when the protein was crystallized. It can also be the case that there are some water molecules locked inside the protein structure, serving a structural role, usually by establishing hydrogen bonds.

If a crystal structure is not available, the next best option is to build a model of the protein, or at least of the active site, using 2D NMR or a homology model. Since these techniques leave room for concern regarding the accuracy of the model, it is not uncommon to use both and compare the resulting models.

## 3.3 2D NMR

Three-dimensional structures can be determined using 2D NMR experiments to find interatomic distances via the nuclear Overhauser effect (NOE). A chemical structure is then determined from a distance–geometry calculation, which optimizes the position of atoms using a force field and the NOE distances as constraints. Usually, this technique is used to create a model of just the active site region. It is often impractical or impossible to measure a sufficient number of NOE distances to produce a good structure for the entire protein.

A number of times in the past, X-ray crystal structures have been determined years after a structure was obtained from NOE measurements. Usually, the

NMR distance–geometry structure has been found to be comparable in accuracy and in agreement to the crystallography results.

## 3.4 HOMOLOGY MODELS

If crystallographic coordinates or a 2D NMR model are not available, then a homology model is usually the next best way of determining the protein structure. A homology model is a three-dimensional protein structure that is built up from fragments of crystallographic models. Thus, the shape of an α-helix may be taken from one crystal structure, the shape of a β-sheet taken from another structure, and loops taken from other structures. These pieces are put together and optimized to give a structure for the complete protein. Often, a few residues are exchanged for similar residues, and some may be optimized from scratch.

Homology models may be very accurate or very marginal, depending upon the degree of identity and similarity that the protein bears to other proteins with known crystal structures. Homology model building is covered in more detail in Chapter 9.

## 3.5 PROTEIN FOLDING

Protein folding is the difficult process of starting with the primary sequence only and running a calculation that tries an incredibly large number of conformers. This is an attempt to compute the correct shape of the protein based on the assumption that the correct shape is the lowest energy conformer. This assumption is not always correct, since some proteins are folded to conformers that are not of the lowest energy with the help of chaperones. It is also difficult to write an algorithm that can determine when disulfide bonds should be formed. Sometimes protein folding gives an accurate model, and sometimes it gives a rather poor model.

The real problem with protein folding is that there is no reliable way to tell whether it has given an accurate model. There are only some checks that provide some circumstantial evidence that the model might be good or might be bad. For example, one can check if hydrophilic residues are on the exterior of the protein and hydrophobic residues are on the interior. Pharmaceutical companies can be justifiably hesitant to spend millions of dollars on research and development based on a folded protein model when there is no way to have confidence in the accuracy of that model. For this reason, protein folding tends to be the last resort for building three-dimensional protein models. Protein folding is discussed in more detail in Chapter 11.

Target identification can be a difficult task, but a vitally important one. Known targets are often ringed with intellectual property rights. Identification of new targets is seen as a means for pharmaceutical companies to stake claim to intellectual turf that will provide them with market share for decades to come.

## BIBLIOGRAPHY

### X-Ray Crystallography

DePristo MA, De Bakker PIW, Blundell TL. Heterogeneity and Inaccuracy in Protein Structures Solved by X-Ray Crystallography. Structure 2004; 12: 831–838.

Drenth J. Principles of Protein X-Ray Crystallography. New York: Springer; 2006.

Ladd MFC, Palmer RA. Structure Determination by X-Ray Crystallography. New York: Springer; 2003.

Woolfson MM. An Introduction to X-Ray Crystallography. Cambridge: Cambridge University Press; 1997.

### 2D NMR

Claridge T. High-Resolution NMR Techniques in Organic Chemistry. Amsterdam: Pergamon; 1999.

Macomber RS. A Complete Introduction to Modern NMR Spectroscopy. New York: Wiley-Interscience; 1998.

### Homology Models

Martz E. Homology Modeling for Beginners with Free Software. Available at http://molvis.sdsc.edu/protexpl/homolmod.htm. Accessed 2008 Nov 17.

### Protein Folding

Zbilut JP, Scheibel T, editors. Protein Folding–Misfolding: Some Current Concepts of Protein Chemistry. Commack: Nova Science; 2006.

Additional references are contained on the accompanying CD.

# 4

# TARGET CHARACTERIZATION

Drug design researchers must come to know the proteins upon which they work—not as members of a "protein socio-political group," but as "biomolecular individuals" with their own likes and dislikes. Knowing the three-dimensional structure of a protein is only the beginning of understanding it. It is also important to understand the mechanism of chemical reactions involving that protein, where it is expressed in the body, the pharmacophoric description, and the mechanism by which inhibitors can bind to it. The following sections of this chapter discuss how this information is obtained.

## 4.1 ANALYSIS OF TARGET MECHANISM

Proteins differ in more than just the shape of the active site. Some proteins are controlled by an allosteric site. Some have an induced fit in the active site, or the whole protein may close to encapsulate the active site. Sometimes, certain spots in the active site are particularly important for specificity. Drug design researchers must understand these nuances in order to design inhibitors for a particular drug target. This includes understanding similarities to other proteins in the same family, as well as differences from the most structurally similar proteins. This understanding comes from a variety of techniques,

*Computational Drug Design.* By David C. Young

both experimental and computational. The following subsections discuss some of the techniques utilized for this purpose.

### 4.1.1 Kinetics and Crystallography

The great value of having X-ray crystallographic coordinates for the target was discussed in the previous chapter. For the purpose of determining the target mechanism, the primary value comes from comparing crystal structures that contain a soaked ligand to the apo structure. If there is an induced fit of the ligand in the active site, this will show any conformational changes that the protein undergoes. If several dissimilar classes of compounds all inhibit the target, crystal structures may be determined with a representative molecule from each group soaked into the crystal. This gives a firm verification that all are binding at the same site, and whether there are any differences in the geometry of the active site.

Measurements of chemical kinetics are also an important experimental tool. Determining whether a reaction exhibits second-order, third-order, or a fractional-order kinetics is one step in understanding its mechanism. Note that proteins in warm-blooded animals have evolved to operate most efficiently in a narrow temperature range. Thus, kinetics can vary significantly if the laboratory's ambient temperature is 20 degrees lower than normal body temperature, although this almost always changes the absolute measurement rather than the reaction order.

### 4.1.2 Automated Crevice Detection

Sometimes, no crystal structure with a soaked inhibitor will be available. When this occurs, it is necessary to determine where the active site is on the protein. One way to do this is to compare the structure with that of a very similar protein. If the two proteins are sufficiently close in structure, then the active site should be in a similar location.

If no experimental data is available to indicate the location of the protein's active site, it is possible to find it computationally. One tool for doing this is software that performs automated crevice detection. These software packages look for concave regions on the protein's surface and categorize them, usually by their volume. This is based on the (usually correct) premise that the largest crevice is most likely to be the active site.

It is wise to verify active site identification by hand. This is usually done by positioning a known inhibitor into each of the likely active sites to verify that the site could indeed hold that inhibitor. It is still possible that a crystal structure including an inhibitor could reveal a surprise not identified by this process; however, that would be a fairly rare circumstance.

### 4.1.3 Transition Structures and Reaction Coordinates

There is very little that can be done experimentally to determine the nuclear positions of all the atoms at every time step as a reaction occurs. Femtosecond spectroscopy comes close to this, but at present is not capable of analyzing entire proteins.

Fortunately, computational chemistry is not bound by the same limitations. Computational techniques are the method of choice for determining entire reaction coordinates. Most analysis of transition structures and reaction coordinates is performed with quantum mechanical calculations, which give the most reliable results. This is not a single calculation, but rather a series of computations. Geometry optimization methods are used to characterize the reactants, products, and van der Waals complexes. Transition state optimization algorithms are used to determine the transition structures. Then intrinsic reaction coordinate (IRC) calculations are used to show the energy and nuclear positions as the molecule traverses the potential energy surface between those points. Other stationary points, such as second-order saddle points, can be taken into account, but it is exceedingly rare that these make any significant contribution to the reaction products or kinetics.

When electron transfer reactions are being modeled, spin density calculations can be carried out to show where the unpaired electrons are at each step of the reaction. Intrinsic reaction coordinate and transition structure calculations can show induced fit behaviors. However, it is up to the researcher to ensure that all the correct motions are being found by the calculation, and not left as negative frequencies.

These techniques come from the small-molecule computational chemistry realm. There are several ways to adapt them to proteins. One option is to work with a truncated model of just the active site. Another option is to use QM/MM techniques, which model the critical atoms with quantum mechanical methods and the rest of the protein with molecular mechanics methods.

### 4.1.4 Molecular Dynamics Simulations

Intrinsic reaction coordinates computed from quantum mechanics give a mathematically well-defined picture of what the reaction may look like. However, in reality, processes at the molecular level are much more complex. Molecules vibrate, but usually not perfectly in just one of the normal vibrational modes. Molecules enter and leave proteins' active sites, but not necessarily following the reaction coordinate exactly. Thus, statistical mechanics and molecular dynamics give a more realistic description of the process, although that description may not be as easily analyzed. Also, these methods are much less suited to describing the bond formation and breaking during the reaction of the native substrate in an enzyme's active site.

One place where molecular dynamics gives useful insight is analyzing the mechanism of proteins that display an induced fit in the active site. Multiple runs through the molecular dynamics simulation of the protein, substrate, and surrounding solvent can give a picture of how the protein structure changes as the substrate enters the active site. It can also provide examples of how the substrate may approach the active site, but fail to undergo a reaction.

A very difficult process to model is an allosteric mechanism in which the substrate causes a change in protein shape, which in turn causes a change in target behavior. An example would be an examination of how taxanes promote tubulin polymerization. Because it is not a competitive inhibition mechanism, docking software is not useful for predicting whether a compound will have the desired antitumor activity. It is necessary to look at multiple molecular dynamics simulations to see if polymerization is enhanced by the presence of a particular drug, as compared with the simulation without a drug.

## 4.2 WHERE THE TARGET IS EXPRESSED

Some proteins are expressed in every cell in the body, while others are expressed only in specific organs. The location in which the drug target is expressed will determine some of the bioavailability concerns that must be addressed in the drug design process. If the target is only expressed in the central nervous system (CNS), then blood–brain barrier permeability must be addressed, either through lipophilicity or through a prodrug approach. Since the blood–brain barrier functions to keep unwanted compounds out of the sensitive CNS, this is a major concern in CNS drug design efforts.

The easiest targets for a drug to reach are cell surface receptors. This is why many drugs are designed to interfere with these receptors, sometimes even when metabolic pathway concerns would suggest that a different target is a better choice. It is not impossible to design a drug to reach a target inside a cell; it simply requires a more delicate lipophilicity balancing act.

## 4.3 PHARMACOPHORE IDENTIFICATION

The pharmacophore is the three-dimensional geometry of interaction features that a molecule must have in order to bind in a protein's active site. These include such features as hydrogen bond donors and acceptors, aromatic groups, and bulky hydrophobic groups. The pharmacophore can be used to search through databases of compounds to identify those that should be assayed. It can also be compared with pharmacophores for other proteins to find other targets that may pose problems with side effects if the drug binds to them. Bioinformatics techniques give an alternative way of finding structurally similar proteins.

Bioinformatics is the more easily used tool, and will find most of the similar proteins. However, there are documented examples of cases where proteins with very different sequences have evolved to perform very similar functions, which could be found by a pharmacophore comparison but not a sequence comparison.

## 4.4 CHOOSING AN INHIBITOR MECHANISM

Most drugs work through a competitive inhibition mechanism. This means that they bind reversibly to the target's active site. While the drug is in the active site, it is impossible for the native substrate to bind. This downregulates the efficiency of the protein, without removing it from the body completely. Competitive inhibitors are the easiest to design with structure-based drug design software packages. They also tend to be the easiest to tune for specificity. Because reversibly bound inhibitors are constantly being cycled through the system, they are also susceptible to being eliminated from the bloodstream quickly by the liver, thus requiring frequent dosages.

Suicide inhibitors bind irreversibly to the target's active site. This makes them inactivators of the target. Completely removing a target from the biological system can have severe side effects. Thus, suicide inhibitors are often used to inactivate proteins in viral and bacterial pathogens, which can be completely removed from the biological system without concerns over side effects.

Uncompetitive inhibitors bind to the enzyme–substrate complex, but not to the free enzyme. This also acts to downregulate protein activity. However, it is more difficult to design uncompetitive inhibitors with specificity for one target. Usually, the resulting complex does not readily dissociate under biological conditions, thus giving uncompetitive inhibitors the disadvantages of suicide inhibitors, without the advantage of specificity. The story is similar for mixed inhibitors (also called noncompetitive inhibitors), which can bind to either the enzyme–substrate complex or the free enzyme.

Allosteric compounds can be uncompetitive inhibitors or noncompetitive inhibitors, or may act as activators to upregulate protein activity. Allosteric regulators may affect the protein in a number of ways. They may upregulate or downregulate protein activity by altering the geometry or vibrational behavior of the protein. Some allosteric compounds act as electron donors or acceptors, or as hydrogen donors or acceptors.

Many issues are involved in selecting targets. These include scientific issues such as those discussed in this text, as well as concerns associated with intellectual property law. A thorough understanding of the target's behavior, mechanism, and role in the metabolic pathway is very useful in guiding drug design efforts. These issues will change the way in which drug design work is approached, which software tools are most appropriate, and how much emphasis is given to each aspect of the work. The following chapters will present typical drug design processes for different types of targets.

## BIBLIOGRAPHY

### IRC

Heidrich D, editor. The Reaction Path in Chemistry: Current Approaches and Perspectives. New York: Springer; 2007.

Young D. Computational Chemistry: A Practical Guide for Applying Techniques to Real World Problems. New York: Wiley; 2001.

### Mechanism

Silverman RB. The Organic Chemistry of Drug Design and Drug Action. 2nd ed. San Diego: Academic Press; 2004.

Voet D, Voet JG. Biochemistry. 3rd ed. New York: Wiley; 2004.

Additional references are contained on the accompanying CD.

# 5

# THE DRUG DESIGN PROCESS FOR A KNOWN PROTEIN TARGET

Not all drugs work in the same way. For example, drugs might compensate for inadequacies in the natural working of the body (e.g., insulin for diabetes), cure a condition (e.g., antibiotics), treat symptoms (e.g., pain killers), alter the patient's emotional state, or prevent pregnancy. Understandably, these different possible types of medications will have to be designed in somewhat different ways. At the design stage, there is an even greater difference based on whether the drug will interact with a protein, DNA, or RNA, or if the target is unknown. This chapter and the next two will discuss different variations on the drug design process that may be encountered in these different situations.

## 5.1 THE STRUCTURE-BASED DESIGN PROCESS

This chapter focuses on drug design procedures generally used when the target protein's structure is known, and the drug is to work through a competitive inhibition mechanism. This design process is called structure-based drug design or alternatively rational drug design or just SBDD. Structure-based drug design is not a single tool or technique. It is a process that incorporates both experimental and computational techniques. This is generally the preferred method of drug design, because it has the highest success rate. This

*Computational Drug Design.* By David C. Young

success is due to the fact that this process gives researchers a clear understanding of how the drug works, and how to improve activity, specificity, pharmacokinetics, etc. The process has been developed to use each tool where it is most effective, while compensating for the shortcomings of specific techniques. However, even with the best tools available, drug design is still a very laborious, difficult task.

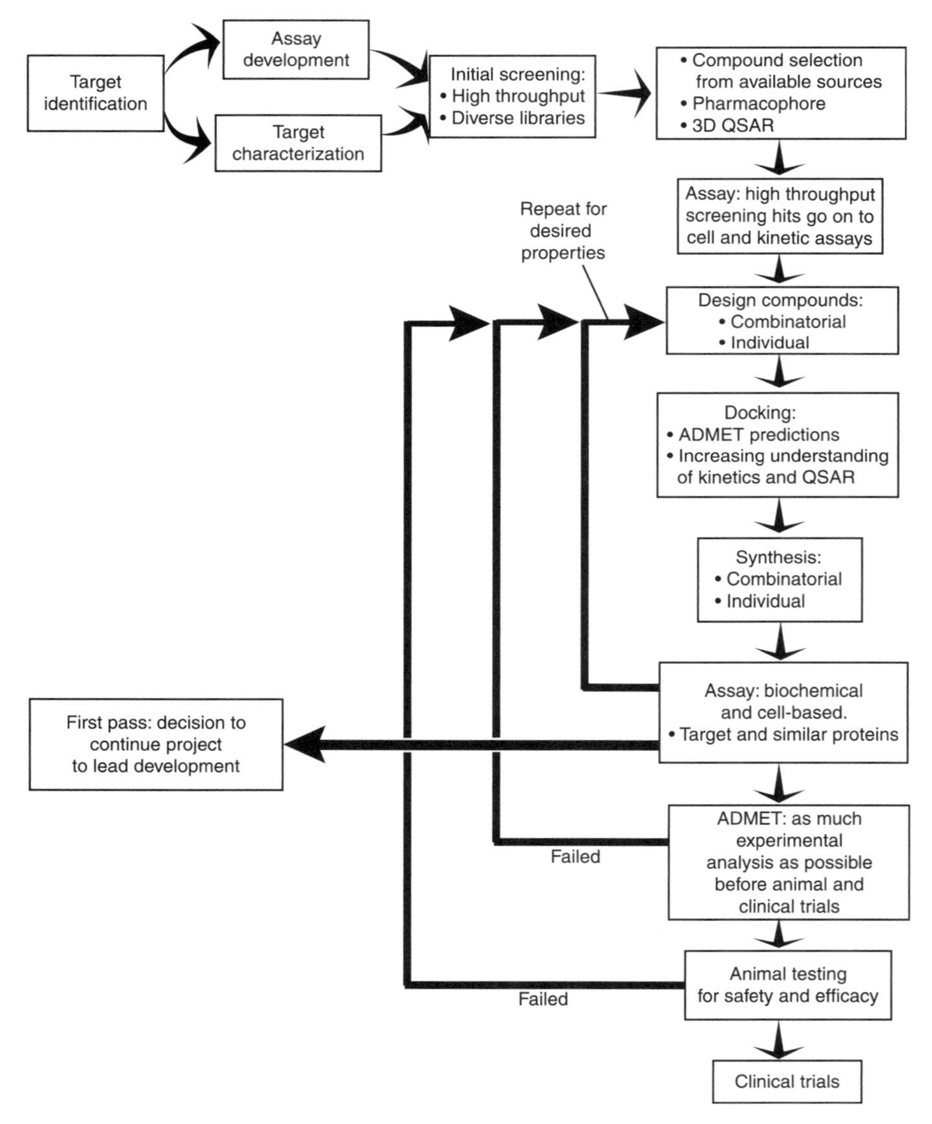


**Figure 5.1** The structure-based drug design process.

Figure 5.1 shows a flow chart of the structure-based drug design process. Some boxes in this figure list multiple techniques for accomplishing the same task. At the target refinement stage, X-ray crystallography is the preferred way to determine protein structures. In the drug design step, docking is the preferred tool for giving a computational prediction of compound activity. The competing techniques may not be used, or may be used only under circumstances where they provide an advantage.

The majority of the work in drug design is aimed at creating compounds with better activity. Finding compounds that can have the desired inhibitory effect requires a significant amount of time, money, and labor. Once a highly active compound is found, it is less difficult to make minor modifications for the purpose of improving toxicity or pharmacokinetics. If it is impossible to achieve the desired oral bioavailability, a prodrug form may be created, as discussed in Chapter 26.

## 5.2 INITIAL HITS

Much of drug design is a refinement process. In this process, successive changes are made to molecular structures in order to improve activity. However, the process needs to get started with some compounds having at least marginal activity. There are often a couple of known inhibitors from previous work on the target, or very similar targets. There often needs to be at least one known inhibitor in order to provide a reference for the development of an assay.

Once an assay has been developed, an initial batch of compounds is assayed. For cost reasons, these are usually compounds that are available commercially, or from previous synthesis efforts. Since the number of commercially available compounds is far too large to assay, it is necessary to select compounds based on some reasonable criteria. There are two approaches that are usually used for this:

- The first approach is to assay a diverse library of compounds that represent many different chemistries. It is expected that an extremely low percentage will be active. However, this has the potential to find a new class of compounds that have not previously been tested for the target being studied.
- The second approach is to search electronic libraries of chemical structures to find those that might fit the active site of the target. This is most often done using a pharmacophore search. Pharmacophore searches have the advantage of not specifying a particular molecular backbone, and of encoding a mathematical description of the geometry of the active site. A 3D-QSAR search is an alternative to a pharmacophore

search, but is typically a better choice when the target geometry is unknown. Pharmacophore and 3D-QSAR searches are discussed in greater detail in Chapters 13 and 15, respectively.

Another computational tool used at this stage is docking. Docking calculations used at this early stage of the drug design process are usually different from the docking calculations used for the main drug design efforts, which are designed for maximum accuracy. There are docking algorithms designed to be extremely fast, at the expense of some accuracy. Because a large quantity of data is being searched, it is necessary to have a technique that takes very little time to analyze each molecule. Chapter 12 is devoted to docking techniques.

A less frequently used tool is a similarity search. Similarity searches can be used to find compounds that are structurally similar to the known inhibitors of the target. Similarity searches tend to find only compounds that are structurally very similar to the known active compounds. They tend not to find compounds with different backbones, but rather only those that contain the same or very similar backbone and functional groups. A substructure search can be used to find compounds with an identical structural motif. These techniques are discussed in greater detail in Chapter 18.

## 5.3 COMPOUND REFINEMENT

The primary advantage of structure-based drug design is that there is a computational way to see what is happening. Docking results allow the researcher to see the inhibitor in the active site. This allows the drug designer to note whether hydrogen bond donors and acceptors are positioned correctly, whether specificity pockets are filled, etc.

Drug designers use a combination of automated compound generation programs and hand-building of compounds in the active site, as shown in Fig. 5.2. This figure follows the way in which a compound may be designed (less all of the dead ends and backtracking). In this hypothetical example, initial screening identified a β-lactam that binds in the active site of the target protein. Frames 2 and 3 show the addition of a carboxylic acid and an amide group to create more points with hydrogen bonds to the active site. Frame 4 shows two methyl groups being added to improve pharmacokinetics and reduce binding to other targets. Frame 5 shows the addition of a phenyl group to further improve oral bioavailability.

The example in Fig. 5.2 is somewhat artificial in that the compound created here is penicillin G, which was created as a derivative of other penicillins, not designed from a β-lactam. Also, in an industrial setting, patent searches would

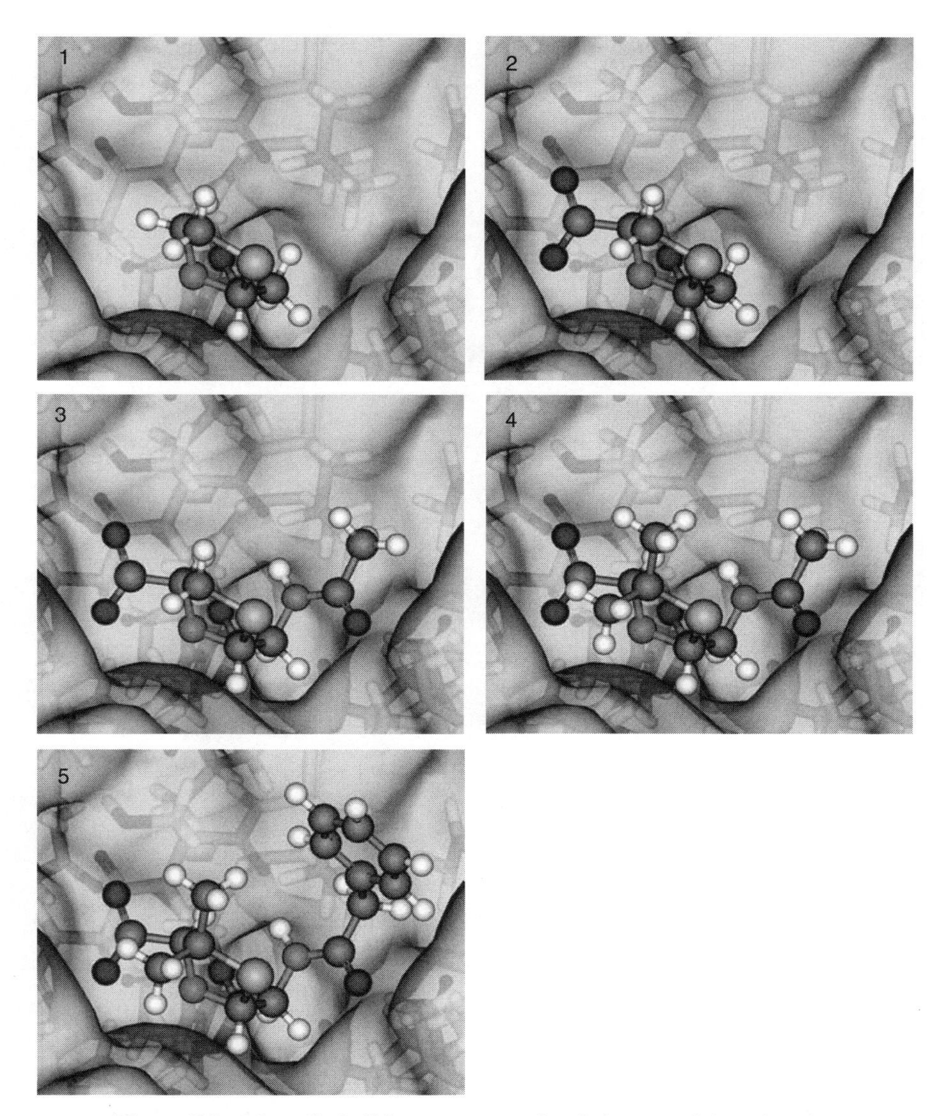


**Figure 5.2** Manually building a compound to fit in a protein's active site.

be conducted to identify that this compound is already patented. Figure 5.3 shows an analysis of the bonding of penicillin G in the active site of the penicillin-binding protein. This type of hand design is still very widely use, in spite of all the automated design tools available. The ability to use software to display the compound in the active site in three dimensions and view it from all angles is very valuable. Figure 5.4 shows the Benchware 3D Explorer from Tripos, which is primarily designed for this type of hand-build work.

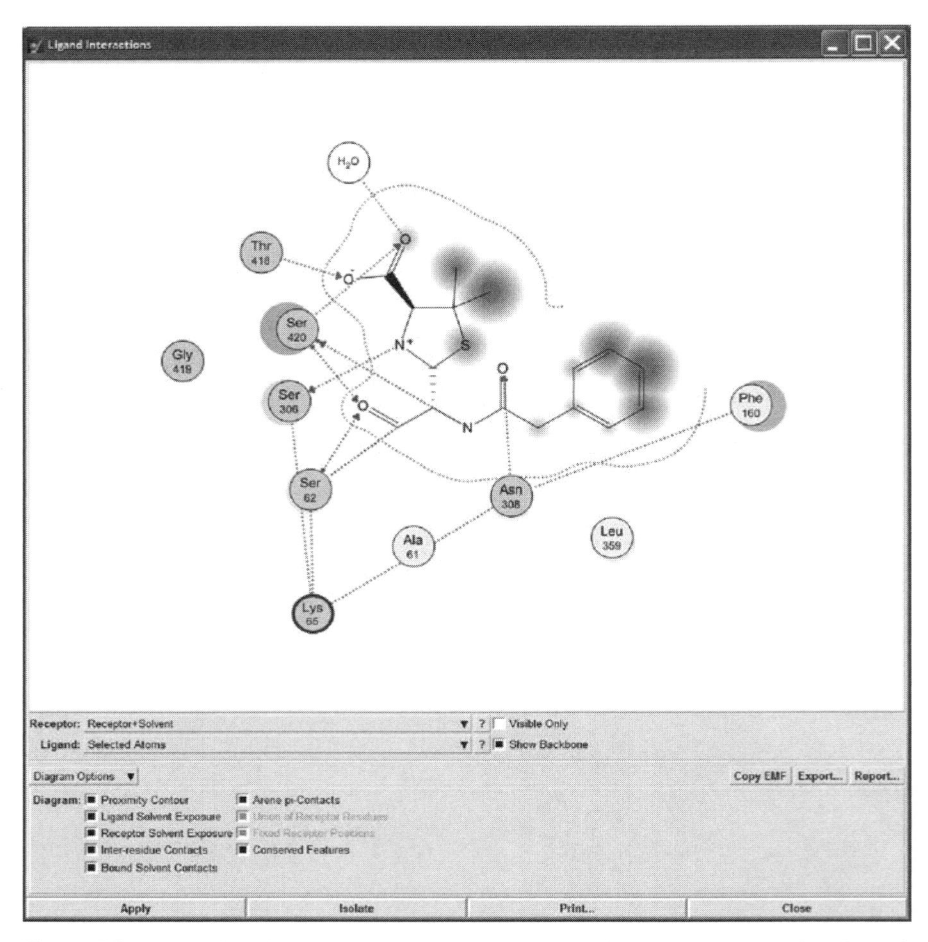


**Figure 5.3** Penicillin G in the active site of the penicillin-binding protein (pdb 2EX8). This display is a ligand interaction analysis generated by the MOE software from Chemical Computing Group.

In doing design work, there are a number of types of alterations to the chemical structure that can be tried to give enhanced activity. In the case of structure-based design, these can first be tried in the computer. The same list is useful in the scenario discussed in the next chapter (unknown target), although that is more of a "blind man's bluff" game. Some alterations that should be considered are:

- bioisosteric substitutions
- variation of substituents (other than bioisosteres)
- extension of the structure

- chain extensions or contractions
- ring expansions or contractions
- trying different rings, or fused ring systems
- simplification of the structure
- making the structure more rigid
- peptidomimetics (substrate analogs or transition state analogs)

Bioisosteric substitution is the process of starting with an initial active compound and replacing functional groups with other groups that have similar biological properties. Unlike isosteres, which have the same number and arrangement of electrons, bioisosteres are selected based on having a similar electronegativity, steric size, lipophilicity, and other properties that affect the

**Figure 5.4** Benchware 3D Explorer from Tripos is a program intended for hand examination of drugs in binding sites and for hand-building. Here regions within reach of hydrogen bond donors and acceptors are displayed as mushroom-shaped regions.

way in which a drug binds to the target. The advantage of using bioisosteric replacement to generate new compound structures is that it is quite likely to create new compounds that are also active. The disadvantage is that any features of the target that were neglected by the initial compound are also neglected by the new compound. Some authors list ring opens and closings as bioisosteric substitutions, while others do not. Some authors give a short list of bioisosteres, sometimes called classical bioisosteres, while others give expanded listings. Figure 5.5 is a listing of bioisosteres. This is an expanded listing compiled from multiple sources. At the extreme end, some researchers have used automated tools to generate lists of tens of thousands of functional group replacements, usually starting from a database of known drugs.

A sometimes-successful type of bioisosterism is retroisosterism. Retroisosterism is the placing of an asymmetric functional group, such as an amide or ester, in the molecule in the reverse order. Figure 5.6 shows an example of retroisosterism.

Variation of substituents is a method of generating many similar compounds, even though the rules of bioisosteric replacement may not be followed. This is often done as a combinatorial library synthesis. A whole host of functional groups may be substituted at each point where the parent compound had a functional group. These focused libraries tend to have a lower hit rate than libraries generated exclusively by bioisosteric replacement. However, they have the advantage of exploring options not suggested by bioisosteric rules. Because of this, a focused library may find a more potent compound that utilizes some feature of the target site not utilized by the original compound.

An extension of the parent structure is usually carried out by adding functional groups in locations where there were none in the parent. This is another way of mapping out structure–activity relationships correlating with substitutions at various points on the parent structure backbone, such as functionalizing rings or chain branching. As with variation of existing substituents, the hit rate is usually lower, but the process may find another way in which the compound can be modified either for improved activity or without loss of activity. Being able to make modifications without loss of activity can be crucial to finding a compound that will have the desired specificity for the intended protein target, without inadvertently inhibiting similar proteins.

Extensions and contractions of both chains and rings are another valuable set of experiments. These may find a derivative with a more optimal fit to the target. There may also be a value in finding compounds with the desired specificity or ADMET ("absorption, distribution, metabolization, excretion, and toxicity") properties.

Different rings or fused ring systems are typically tried, both to find a more optimal fit to the target and to make the structure more rigid. The majority of

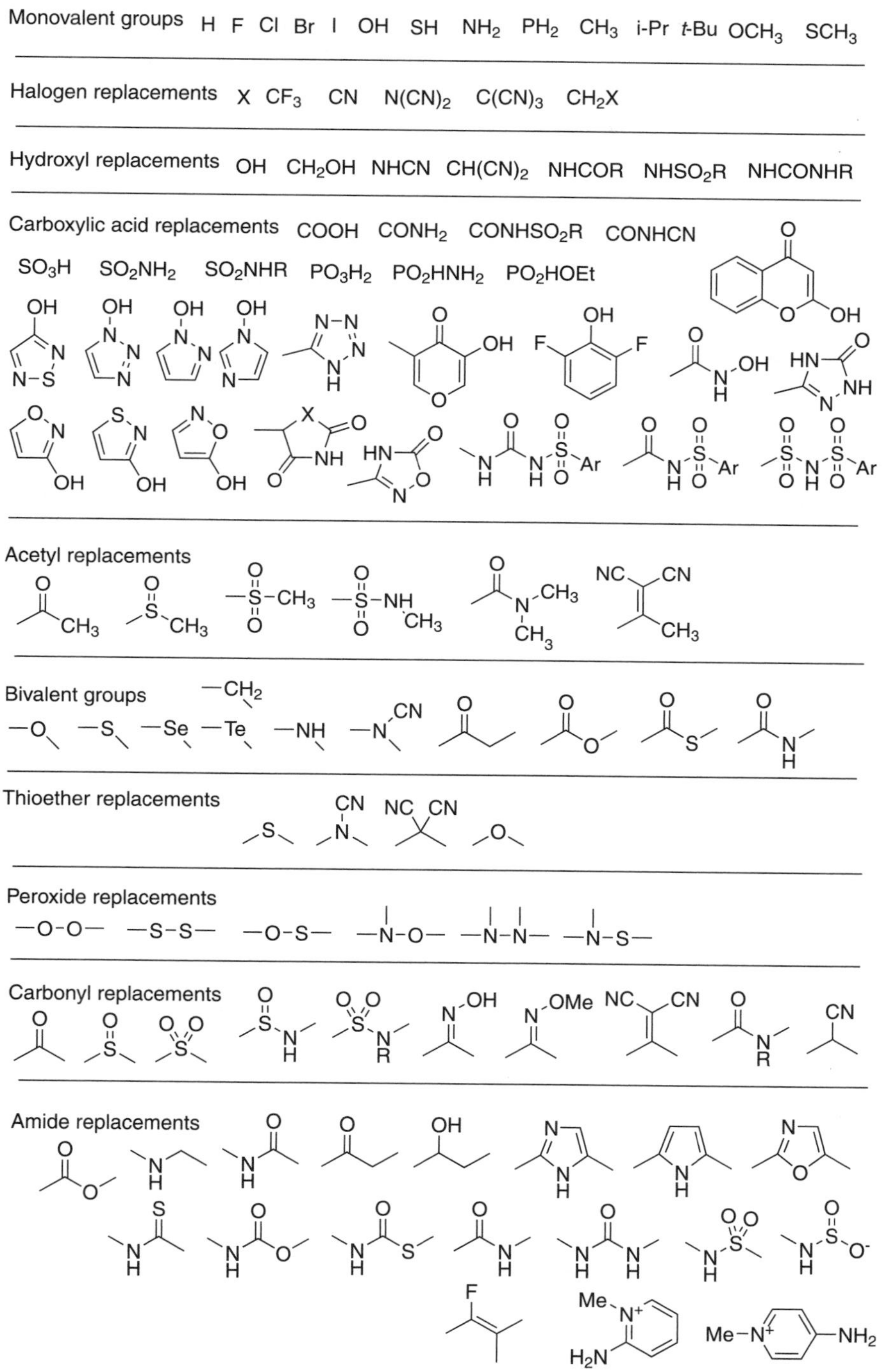

Monovalent groups H F Cl Br I OH SH $NH_2$ $PH_2$ $CH_3$ i-Pr *t*-Bu $OCH_3$ $SCH_3$

Halogen replacements X $CF_3$ CN $N(CN)_2$ $C(CN)_3$ $CH_2X$

Hydroxyl replacements OH $CH_2OH$ NHCN $CH(CN)_2$ NHCOR $NHSO_2R$ NHCONHR

Carboxylic acid replacements COOH $CONH_2$ $CONHSO_2R$ CONHCN $SO_3H$ $SO_2NH_2$ $SO_2NHR$ $PO_3H_2$ $PO_2HNH_2$ $PO_2HOEt$

Acetyl replacements

Bivalent groups

Thioether replacements

Peroxide replacements —O-O— —S-S— —O-S— —N-O— —N-N— —N-S—

Carbonyl replacements

Amide replacements

**Figure 5.5** Bioisosteres.

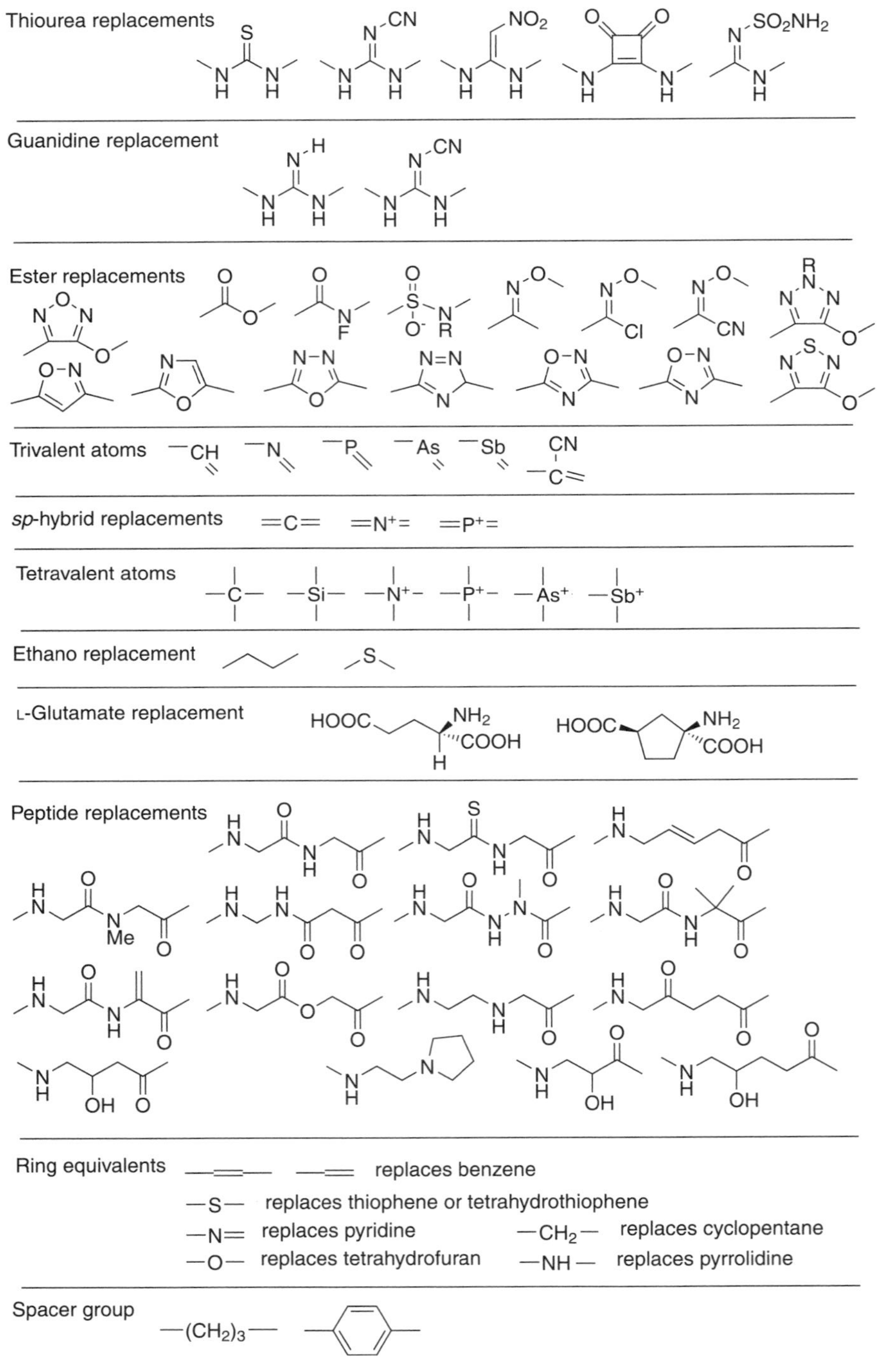

Thiourea replacements

Guanidine replacement

Ester replacements

Trivalent atoms —CH —N —P —As —Sb CN —C=

*sp*-hybrid replacements =C= $=N^+=$ $=P^+=$

Tetravalent atoms —C— —Si— $—N^+—$ $—P^+—$ $—As^+—$ $—Sb^+—$

Ethano replacement

L-Glutamate replacement

Peptide replacements

Ring equivalents

—S— replaces thiophene or tetrahydrothiophene

—N= replaces pyridine

$—CH_2—$ replaces cyclopentane

—O— replaces tetrahydrofuran

—NH— replaces pyrrolidine

Spacer group $—(CH_2)_3—$

**Figure 5.5** *(Continued)*

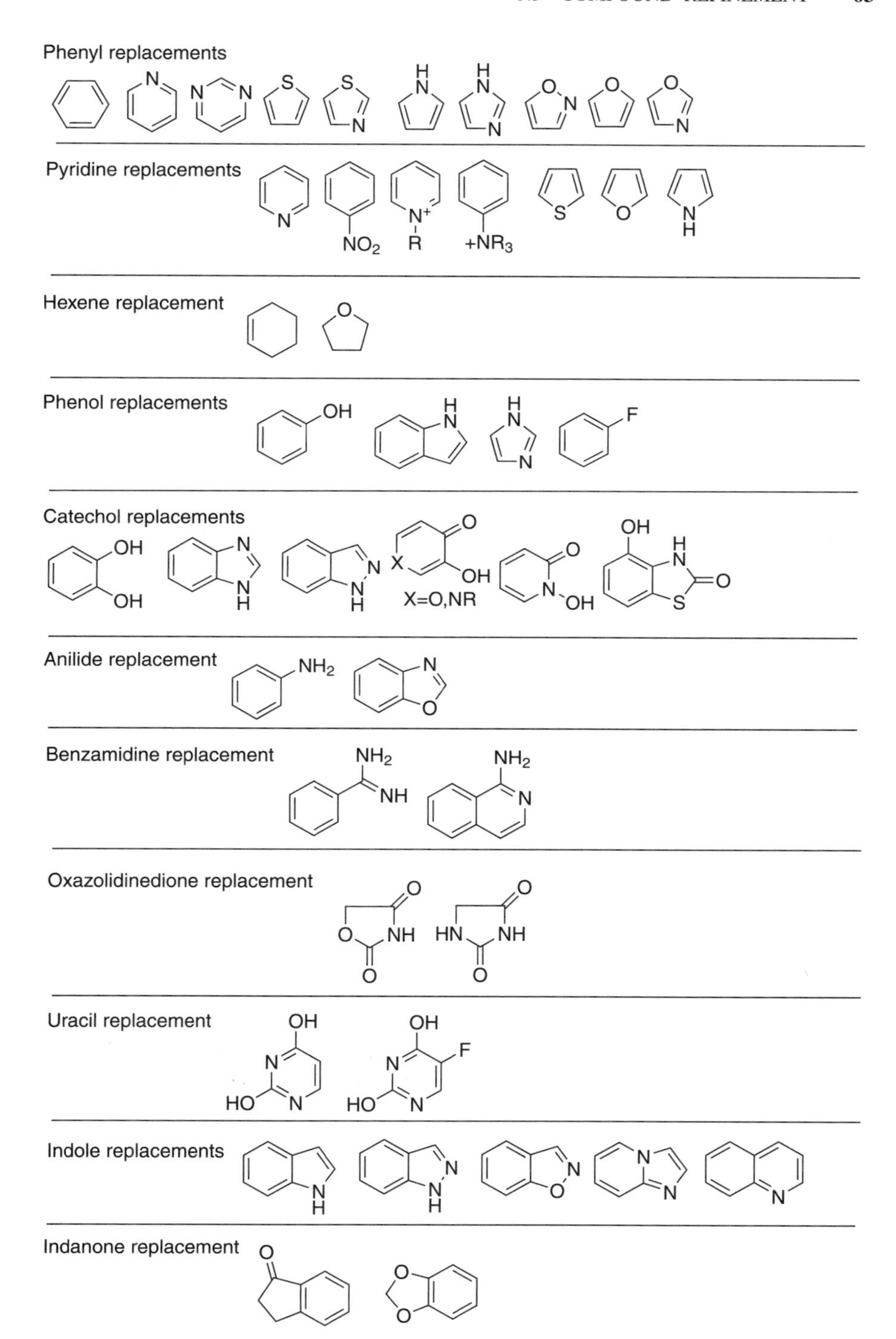


**Figure 5.5** *(Continued)*

Heterocycle replacements
NC-switch

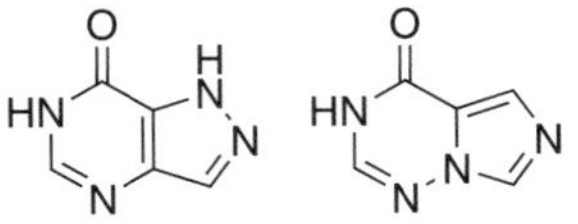

Naphthyl replacement
ring opening

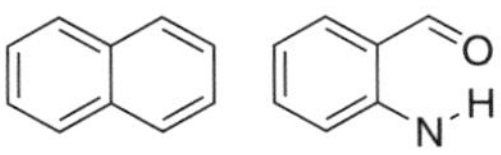

Carbazole replacements

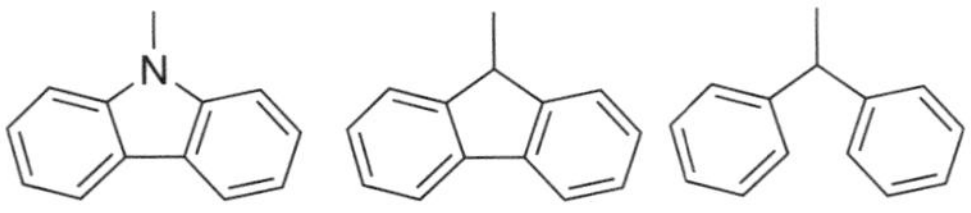

Glipine replacements

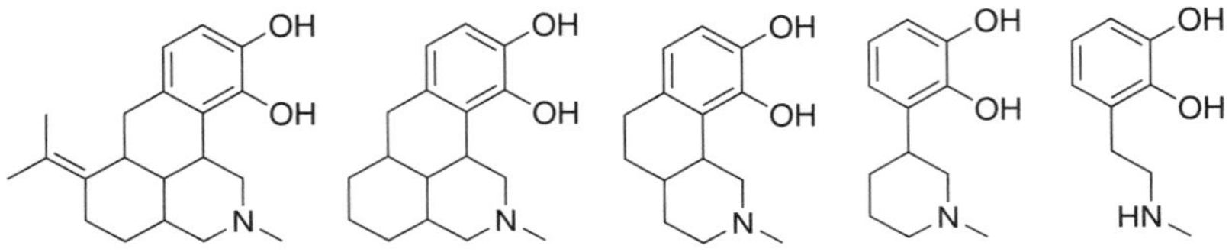

GABA agonist replacements

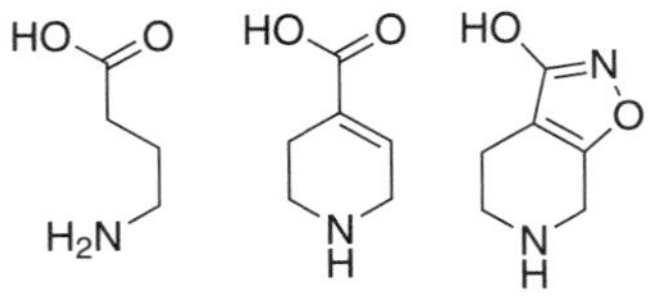

Clonidine replacement

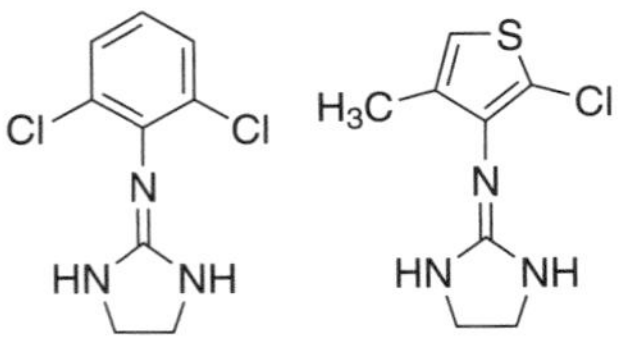

Estradiol replacement

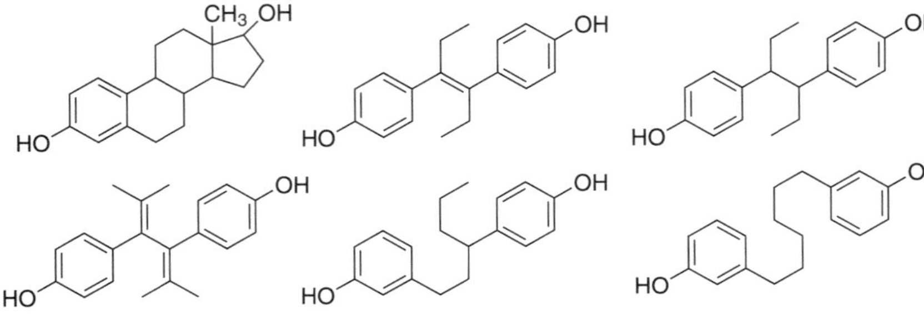

**Figure 5.5** *(Continued)*

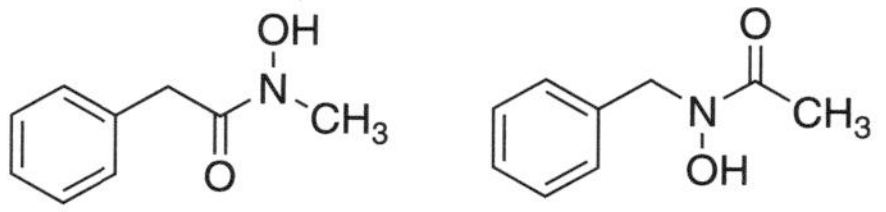

**Figure 5.6** An example of retroisosterism, a form of bioisosterism in which an asymmetric functional group is placed in the molecule in reverse order.

drugs tend to have heterocyclic, fused ring systems. Making a structure more rigid improves the statistical likelihood of binding to the active site on any given collision event owing to the structure being frozen in the correct conformation. Fixing the structure in the conformation for one target also prevents it from taking on a conformation that will allow it to bind to an unintended protein. Changing the ring structure could help or hurt the ADMET properties.

Drug designers can fall into a mental trap of always making structures more complex in order to get a more optimal fit to the target's active site. However, there are times when it may be better to have a simpler structure. Smaller, simpler structures tend to pass through the blood–brain barrier more readily. Simpler structures can also be better for antibiotics and antivirals. If antibiotics or antivirals are made very complex in order to fit perfectly in the active site of a pathogen protein, then only a very small mutation of the pathogen could be necessary to generate a serotype that is resistant to the drug. This is often checked as much as possible by assaying the drug against multiple serotypes of the pathogen. Simpler structures can also be less expensive to manufacture, which is rarely a concern in creating drugs for humans, but may be an important design criterion for designing agrochemicals, which must be reasonably priced in order to generate sufficient sales volumes.

The native substrate for many drug targets is a peptide. However, peptides make poor drugs, as they are too readily metabolized and excreted, and often bind to multiple receptors. Thus, one strategy for drug design is to use the native substrate as a template for designing a peptidomimetic, a nonpeptide compound that has a similar shape and binding to the active site. A key part of this process is replacing amide groups with similar shaped groups that will not be cleaved by proteases. Other groups that are readily broken down in the body, such as esters, are also generally avoided. Likewise, if the native substrate is a nucleotide or carbohydrate, then the drug can be designed to mimic those species. Chapter 2 contains further discussion of substrate analogs and transition state analogs, both of which are often peptidomimetics.

These structural modifications are often made both in structure-based drug design and when the geometry of the active site is not known. When the site is unknown, there is a greater need to try every conceivable permutation on the structure. When the active site is known, some of the modifications can be clearly ruled out by computational docking analysis of how the compound will (or will not) fit in the active site.

Once molecules have been designed for improved drug activity, they must be tested. The first round of testing is in the computer. The most heavily used tool for this testing is docking (Chapter 12). Docking programs capture all of the energy contributions that are used in molecular mechanics. Some docking programs have the ability to allow for solvation, induced fit, entropy corrections, and, in rare cases, inhibitors being covalently bound to the active site. The docking results show the optimal molecular position in the active site, and report a quantitative binding energy.

Sometimes, pharmacophore models and 3D-QSAR models are also used to test the activity of designed compounds. However, these techniques tend to give qualitative pass/fail results, instead of the quantitative results given by docking programs. This is why docking is the most heavily used computational technique for testing compounds in structure-based drug design. Pharmacophore and 3D-QSAR models are used more heavily when the target is unknown, and it is thus impossible to use docking. Pharmacophore models are often the method of choice for searching large databases of available compounds to find those that should be screened.

Docking calculations are also typically performed to determine whether the drug will also bind to other proteins in the body, particularly those that are most similar to the target. A careful analysis of these other proteins can reveal regions of difference in the active site, so that a functional group can be placed on the drug to give better specificity. Specificity is desired to minimize the potential for unwanted side effects. The inclusion of chiral centers in the drug can also improve specificity.

A number of software packages have been created to automatically design compounds to fit in the target active site. Packages called *de novo* packages use some form of artificial intelligence algorithm to accomplish this. Some *de novo* programs dock a large number of simple molecular fragments into the active site, and then connect them together to build molecules. Others build up a compound from a starting group. Drug design can also be done using software packages for designing combinatorial libraries. *De novo* programs are discussed further in Chapter 17, and combinatorial library design in Chapter 8.

Once compounds have been tested computationally, they must be synthesized, and then tested in the laboratory. This is done with biochemical and cell-based assays. Typically, a dilution series is tested at this stage of the design process in order to determine the binding kinetics. Assays against similar proteins or pathogen serotypes may also be carried out. If ADMET issues are expected to be a key issue for the drug being designed (e.g., drugs to treat central nervous system disorders), then ADMET may be tested both computationally and experimentally in cell cultures at this stage of the drug design.

## 5.4 ADMET

In addition to designing drugs for high activity, drug designers must also be aware of concerns over absorption, distribution, metabolization, elimination, and toxicity (ADMET). There are software packages for predicting ADMET properties. These software packages are not yet very accurate. However, the potential cost of failed clinical trials is so high that such concerns must be considered, in any way possible. In the compound refinement stage of development, these issues do not prevent researchers from exploring certain chemistries in the quest for high activity. However, they may be used later to prioritize which compounds go to animal trials first.

There are cases where the structure of drugs is directly influenced by ADMET concerns. For example, functional groups may be added to the drug backbone in a position where there is little effect on drug binding, purely in order to make the drug more or less lipophilic to improve bioavailability.

Another technique is metabolic blocking. For example, adding a methyl group next to a *cis* carbon–carbon bond can prevent that bond from being readily oxidized (which in turn allows the drug to be eliminated from the body more rapidly). Functional groups that can be readily metabolized, such as methyl groups on aromatic rings, may also be removed from the structure to slow down metabolism of the drug. ADMET prediction software is discussed in Chapter 19.

## 5.5 DRUG RESISTANCE

Drug resistance is an issue of great concern in medicine. The rise of drug-resistant strains gives antibiotics and antivirals a limited useful life. In order to slow the emergence of drug-resistant strains, physicians are encouraged to prescribe these treatments sparingly. Of even greater concern is the emergence of multidrug-resistant strains of some particularly virulent pathogens, such as multidrug-resistant methicillin-resistant *Staphylococcus aureus* (MRSA).

There are six known mechanisms by which pathogenic viruses and bacteria build up resistance to drugs:

- *A mutation of the organism may result in a change in the shape of the drug target's active site.* When this happens, the drug may no longer fit in the active site, with the result that the organism is resistant to the drug. This has been identified as the mechanism behind the increased resistance to tetracycline. This possibility leads to deciding upon a trade-off during the drug design process. An antibiotic that fits perfectly in the active site may allow the organism to build up resistance very quickly, where as an antibiotic that fits too loosely in the active site may have more side effects due to interactions with other targets in the body.

- *Some bacteria contain enzymes that break down foreign compounds.* For example, bacterial β-lactamases break down penicillins. In this case, the mechanism of breakdown should be examined. The enzymes responsible for breaking down foreign compounds may be upregulated or undergo mutations that make an organism more efficient at destroying the drug. It may be possible to design the drug to avoid, or slow down, the rate at which the pathogen can destroy it by making it a poor fit to these enzymes.
- *There can be upregulation of the drug efflux from the bacterial cell.* This is simply an increase in the production of efflux pumps already existing in the bacterial genome. This gives increased resistance to drugs, but also causes the bacteria to grow and reproduce more slowly owing to the diversion of energy in the form of ATP to the needs of the efflux pump. This has been identified as a factor in the increased resistance of bacteria to tetracycline and quinolone antibiotics.
- *Bacteria can undergo genetic changes that result in decreased cell wall permeability.* This has been identified as a mechanism behind the rise of increasingly drug-resistant strains of tuberculosis.
- *There can be changes in the pathogen's biochemical pathways.* Over time, pathogens can develop new biochemical pathways. When this happens, the pathway that is targeted by a drug goes from being crucial for the organism's survival to being unused or redundant.
- *Some bacterial colonies can excrete carbohydrates that join to form protective biofilms.* These biofilms can inhibit the ability of drugs to reach the intended target.

This chapter has presented the structure-based drug design process and the tools utilized in that process. These topics are of very great importance to drug designers. This is because these are the most heavily used, and most successful, of the computational drug design techniques. The next two chapters will examine what can be done when structure-based drug design cannot be used, or must use a modified format due to an atypical drug target.

## BIBLIOGRAPHY

### Structure-Based Drug Design

al-Obeidi F, Hruby VJ, Sawyer TK. Peptide and peptidomimetic libraries. Molecular diversity and drug design. Mol Biotechnol 1998; 9: 205–223.

Balbes LM, Mascarella SW, Boyd DB. A perspective of modern methods in computer-aided drug design. In: Lipkowitz KB, Boyd DB, editors. Reviews in Computational Chemistry, Volume V. New York: VCH; 1994: 337–379.

Goodman M, Shao H. Peptidomimetic building blocks for drug discovery: An overview. Pure Appl Chem 1996; 68: 1303–1308.

Parrill AL, Reddy MR, editors. Rational Drug Design: Novel Methodology and Practical Applications. Washington: American Chemical Society; 1999.

Patrick GL. An Introduction to Medicinal Chemistry. Oxford: Oxford University Press; 1995.

Truhlar DG, Howe WJ, Hopfinger AJ, Blaney J, Dammkoehler RA, editors. Rational Drug Design. New York: Springer; 1999.

### Bioisosteres

Herdewijn P, De Clercq E. The cyclohexene ring as a bioisostere of a furanose ring: Synthesis and antiviral activity of cyclohexenyl nucleosides. Bioorg Med Chem Lett 2001; 11: 1591–1597.

Lima LM, Barreiro EJ. Bioisosterism: A useful strategy for molecular modification and drug design. Curr Med Chem 2005; 12: 23–49.

Lin LS, Lanza TJ Jr, Castonguay LA, Kamenecka T, McCauley E, Van Riper G, Egger LA, Mumford RA, Tong X, MacCoss M, Schmidt JA, Hagmann WK. Bioisosteric replacement of anilide with benzoxazole: Potent and orally bioavailable antagonists of VLA-4. Cheminformatics 2004; 35: 2331–2334.

Molecular Topology. SiBIS (Similarity in BioIsosteric Space) covers Bioisostere. Available at http://www.moltop.com/science_bioisostere.htm. Accessed 2008 Feb 22.

Nogrady T, Weaver DF. Medicinal Chemistry; A Molecular and Biochemical Approach. Oxford: Oxford University Press; 2005.

Pandit NK. Introduction to the Pharmaceutical Sciences. Philadelphia: Lippincott Williams & Wilkins; 2006.

Schneider G, Baringhaus KH. Molecular Design. Weinheim: Wiley-VCH; 2008.

Patani GA, LaVoie EJ. Bioisosterism: A Rational Approach in Drug Design. Chem Rev 1996; 96: 3147–3176.

Silverman RB. The Organic Chemistry of Drug Design and Drug Action. London: Academic Press; 2004.

Stewart KD, Shiroda M, James CA. Drug Guru: A computer software program for drug design using medicinal chemistry rules. Bioorg Med Chem 2006; 14: 7011–7022.

## SUCCESS OF STRUCTURE-BASED DRUG DESIGN

Huang Z. Drug Discovery Research: New Frontiers in the Post-Genomic Era. Hoboken, NJ: Wiley-Interscience; 2007.

National Institute of General Medical Sciences. Structure-Based Drug Design Fact Sheet: NIGMS-Supported Structure-Based Drug Design Saves Lives. Available at http://www.nigms.nih.gov/Publications/structure_drugs.htm. Accessed 2008 Oct 3.

### Drug Resistance

Centers for Disease Control and Prevention. Antibiotic/Antimicrobial Drug Resistance Home Page. Available at http://www.cdc.gov/drugresistance. Accessed 2008 Feb 20.

Corey EJ, Czakó B, Kürti L. Modules and Medicine. Hoboken: Wiley; 2007.

McKie JH, Douglas KT, Chan C, Roser SA, Yates R, Read M, Hyde JE, Dascombe MJ, Yuthavong Y, Sirawaraporn W. Rational drug design approach for overcoming drug resistance: Application to pyrimethamine resistance in malaria. J Med Chem 1998; 41: 1367–1370.

Varghese JN, Smith PW, Sollis SL, Blick TJ, Sahasrabudhe A, McKimm-Breschkin JL, Colman PM. Drug design against a shifting target: A structural basis for resistance to inhibitors in a variant of influenza virus neuraminidase. Structure 1998; 6: 735–746.

Yin P, Das D, Mitsuya H. Overcoming HIV drug resistance through rational drug design based on molecular, biochemical, and structural profiles of HIV resistance. Cell Mol Life Sci 2006; 63: 1706–1724.

Additional references are contained on the accompanying CD.

# 6

# THE DRUG DESIGN PROCESS FOR AN UNKNOWN TARGET

## 6.1 THE LIGAND-BASED DESIGN PROCESS

There are times when a drug's target protein is not known. This is still fairly common, although it is slowly becoming less so as the body of knowledge about biological systems expands. Today, it is sometimes the case that the target is not known at the beginning of a project, but is determined part way through development. The odds of success are greater if the target is known and a structure-based drug design process can be followed. However, there are times when there is a good business reason for pursuing the design of a drug without a known target. For example, cell surface receptors make excellent drug targets, but are very difficult to crystallize. In recent years, the FDA has requested more information about drugs' mechanisms of action, which could portend the permanent obsolescence of the design process presented in this chapter.

Some of the techniques used for structure-based drug design, such as docking and biochemical assays, cannot be used if the target structure is not known. This chapter presents the process for designing drugs without a known target structure: ligand-based drug design (LBDD)—see Fig. 6.1.

*Computational Drug Design*. By David C. Young

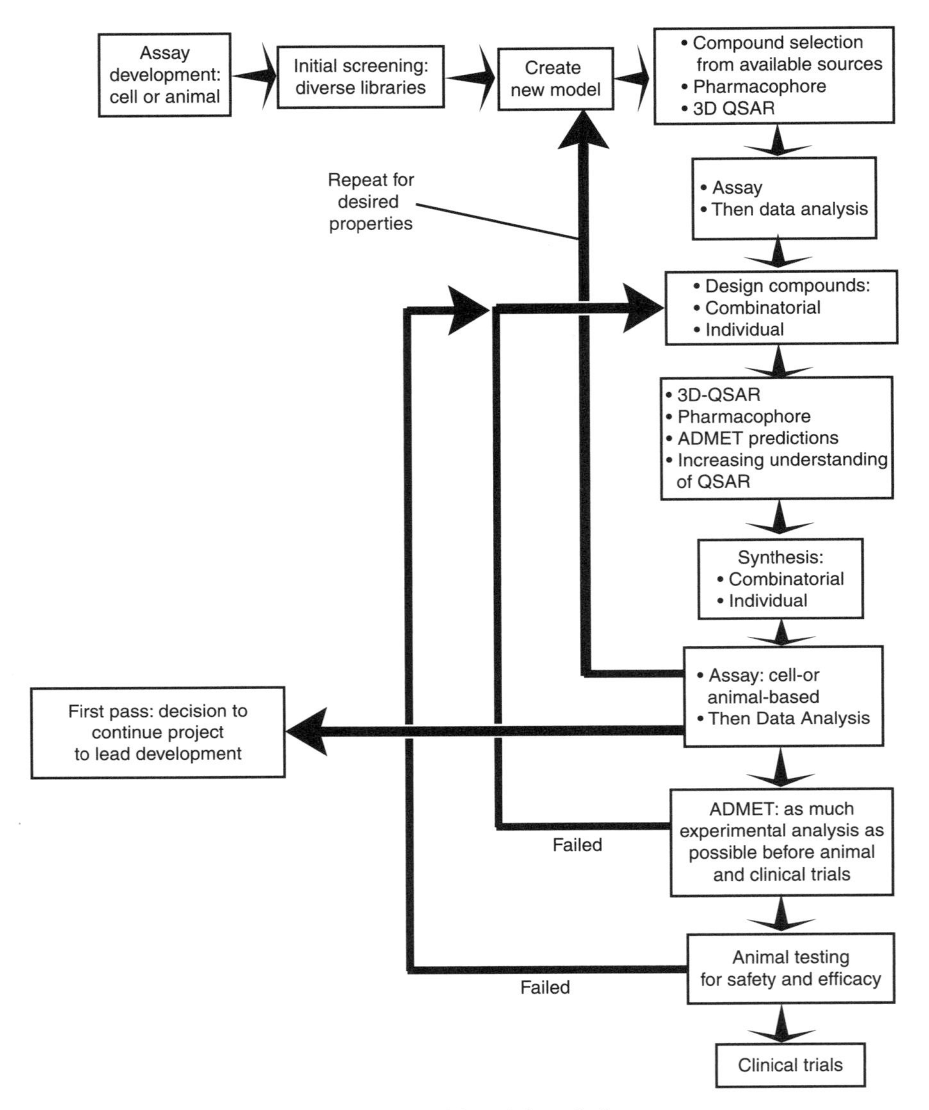


**Figure 6.1** The ligand-based drug design process.

## 6.2 INITIAL HITS

For any type of drug design effort, there needs to be a way to test if trial compounds have activity. In the case where the target is unknown, this must be done preferably through a cell culture assay, or with an animal test as a last resort. Cell culture tests are preferred for cost and convenience reasons.

Once the assay is developed, compounds must be tested to find ones that are active. If there are already known active compounds, similar compounds can be tested. If there are very few known active compounds, the best option is to also test a diverse library of compounds. There should be at least one known active compound in order to validate that the assay works correctly. Many times, a diverse library is screened even if there are known active compounds. This is done in an attempt to find new classes of compounds that are active, which is referred to as "lead hopping."

## 6.3 COMPOUND REFINEMENT

A collection of active compounds can be used to develop computational models that do not require knowledge of the geometry of the active site. These include pharmacophore models (Chapter 13) and 3D-QSAR models (Chapter 15). 3D-QSAR is the slightly more prevalent technique, owing to its sometimes better accuracy. It tends to be used more for computational testing of compounds that have been designed computationally. Pharmacophore models tend to be used more heavily for searching large databases of existing compounds, as pharmacophore searches execute much more quickly.

Other than 3D-QSAR and pharmacophores in place of docking, compound refinement is still an iterative process of designing compounds, testing them computationally, synthesis, and experimental testing. The design process is more trial and error, since it is not possible to examine how a compound fits in the active site. Considerable effort is often expended synthesizing and testing compounds with a functional group at each point on a known active compound, testing all stereoisomers, etc. It is also typical that the 3D-QSAR and pharmacophore models are refined multiple times as more assay data becomes available.

The analog to hand-building of compounds in structure-based drug design is the use of structure–activity relationships (SAR). These are qualitative relationships that find correlations between activity and structure, such as suggesting that a hydrogen bond acceptor should be in a particular position in the drug structure. SAR tools are discussed further in Chapters 8 and 18.

After creating or refining the models, an additional round of testing of existing compounds is usually carried out. This is done because it is more cost-effective to test compounds that are already available than to synthesize all compounds from scratch. The molecule selection criterion on this pass tends to be very tight, in order to identify only compounds that fit the model closely. This focused search is sometimes referred to as "cherry picking."

There are a number of screening data analysis tools that are designed to help mine SAR and other useful information out of the screening data. These tools are used in both structure-based and ligand-based drug design processes, but they are perhaps more critical to success in a ligand-based drug design process. They are discussed in more detail in Chapter 18.

## 6.4 ADMET

Absorption, distribution, metabolization, elimination, and toxicity (ADMET) analysis in ligand-based drug design differs only slightly from the corresponding analysis in structure-based drug design. The major difference is that researchers do not necessarily know what organ the target is expressed in. In the case of an unknown target, the researchers cannot be sure if the compound being created by trial and error is the active form or a prodrug form, although the necessity for ester or amide groups can be a hint that it is a prodrug form.

Drug design for unknown targets may eventually become a thing of the past. However, until then, there will still be targets for which there is an economic incentive to pursue such work, in spite of the lower probability of success.

## BIBLIOGRAPHY

Bacilieri M, Moro L. Ligand-based drug design methodologies in drug discovery process: An overview. Curr Drug Discov Technol 2006; 3: 155–165.

Jahnke W, Erlanson DA, Mannhold R, Kubinyi H, Folkers G, editors. Fragment-Based Approaches in Drug Discovery. Weinheim: Wiley-VCH; 2006.

Moro S, Bacilieri M, Deflorian F. Combining ligand-based and structure-based drug design in the virtual screening arena. Expert Opin Drug Discov 2007; 2: 37–49.

Rognan D, Mannhold R, Kubinyi H, Folkers G, editors. Ligand Design for G Protein-Coupled Receptors. Weinheim: Wiley-VCH; 2006.

Silverman RB. Organic Chemistry of Drug Design and Drug Action. San Diego: Academic Press; 2004.

Additional references are contained on the accompanying CD.

# 7

# DRUG DESIGN FOR OTHER TARGETS

Most of the discussion in the last two chapters was focused on the most typical drug design processes. As such, most of that information is relevant to designing drugs that will reversibly bind to proteins and act through competitive inhibition. When designing drugs for an unknown target, researchers tend to think in terms of designing a competitive inhibitor—for lack of data to the contrary, and because statistically this is the most likely scenario. Competitive inhibition of protein targets is the mode of action of the majority of drugs, but there are a sizable number of drug design efforts that do not fit this description. Schneider and Baringhaus report an analysis of the targets for 8500 drugs and lead compounds, which is summarized in Table 7.1. In contrast to these results, Filmore estimates that half of all drugs interact with G-protein-coupled receptors. This chapter discusses some drug design issues that arise in situations other than competitive inhibition of proteins.

**TABLE 7.1 Type of Drug Targets for Drugs and Lead Structures**

| Target Type | Percentage |
|---|---|
| Protease | 19 |
| G-protein-coupled receptor | 16 |
| Ion channel | 13 |
| Kinase | 12 |
| Nuclear receptor | 4 |
| Other enzymes | 17 |
| Other nonenzyme targets | 19 |

## 7.1 DNA BINDING

Only a small percentage of drugs work by binding to DNA. The most well known of these are cancer chemotherapy compounds, such as cisplatin and carboplatin. Some antibacterial and antiviral drugs also bind to DNA. The majority of these drugs bind in the minor groove of the DNA strand, as shown in Fig. 7.1. An affinity for binding to AT-rich sequences, where the minor groove is a narrow crevice, is not unusual. Some small-molecule drugs act through intercalation between base pairs, as shown in Fig. 7.2.

The grooves in a DNA strand are not as deep and well defined as is typical of the active sites of proteins. Thus, the tight "lock-and-key" fit inherent in protein inhibitors is not present. Also, similar DNA sequences may occur at multiple places in the genome. Thus, it may not be possible to optimize for specificity by fitting to the active site. A compound can be designed to have better binding near certain DNA sequences, but will often have somewhat weaker binding to similar sequences. The functional groups that face away from the DNA strand when the compound is bound can be tuned to improve pharmacokinetics and minimize accidental binding in the active sites of proteins. Because the places where the compound may be incorrectly binding may be unknown, this is more of a trial-and-error process than is structure-based drug design. Drugs that bind to DNA through intercalation tend to have even less sequence specificity.

The specificity concerns raised in the previous paragraph lead one to question why drugs that bind to DNA can work at all. The primary reason that these drugs do work is that cancer cells and pathogenic organisms tend to have a higher metabolic rate than other cells in the body. Thus, more of the drug affects the target cells purely because they take it up faster than the other cells in the body, and reproduce faster. Thus, the only way to keep side effects down is to ensure that the drug stays in the body long enough to be taken up by the target cells, but not long enough to have a significant effect on other cells.

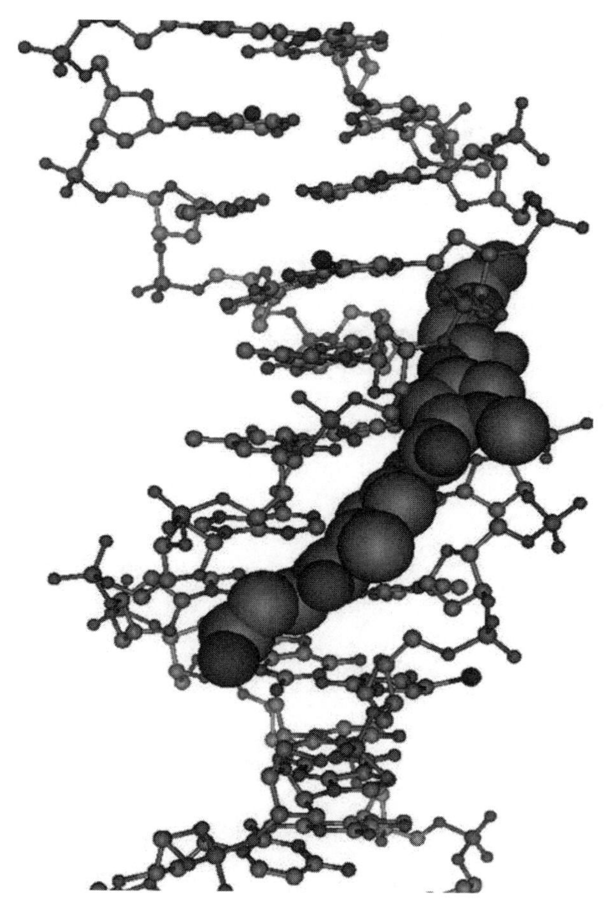

**Figure 7.1** Netropsin binding in the minor groove of a DNA strand (pdb 101D).

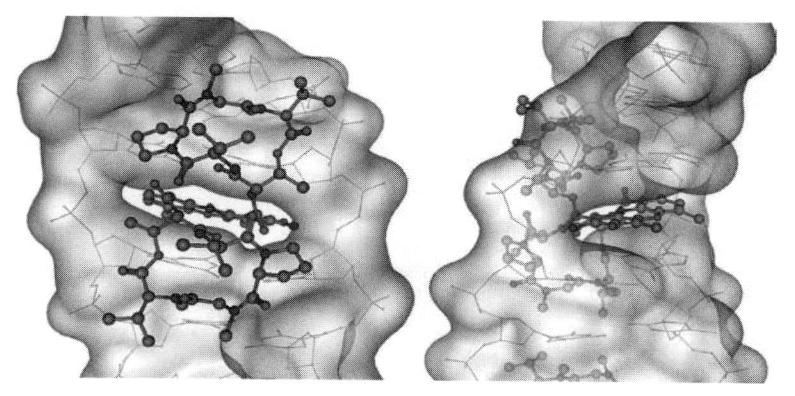

**Figure 7.2** Front and back views of dactinomycin (actinomycin D) intercalated between the base pairs of a DNA strand (pdb 1MNV).

## 7.2 RNA AS A TARGET

Very few drugs bind to RNA, but there are some examples. Drugs that bind to RNA may be wrapped around the RNA strand, or may be encapsulated in a pocket in the folded RNA. One of the best known RNA-binding drugs is streptomycin. Its antibiotic activity is due to binding to the RNA of pathogenic bacteria, such as tuberculosis. This inhibits protein synthesis in the bacteria, which in turn damages the cell membrane. Figure 7.3 shows an example of streptomycin binding to RNA.

Computational design of drugs similar to streptomycin is, in theory, similar to design of protein inhibitors. However, in practice, it poses some technical difficulties. Docking programs are designed to test positions of molecules that are mostly encapsulated in the active site of a protein. These algorithms do not always work as well when the inhibitor is wrapped around the target.

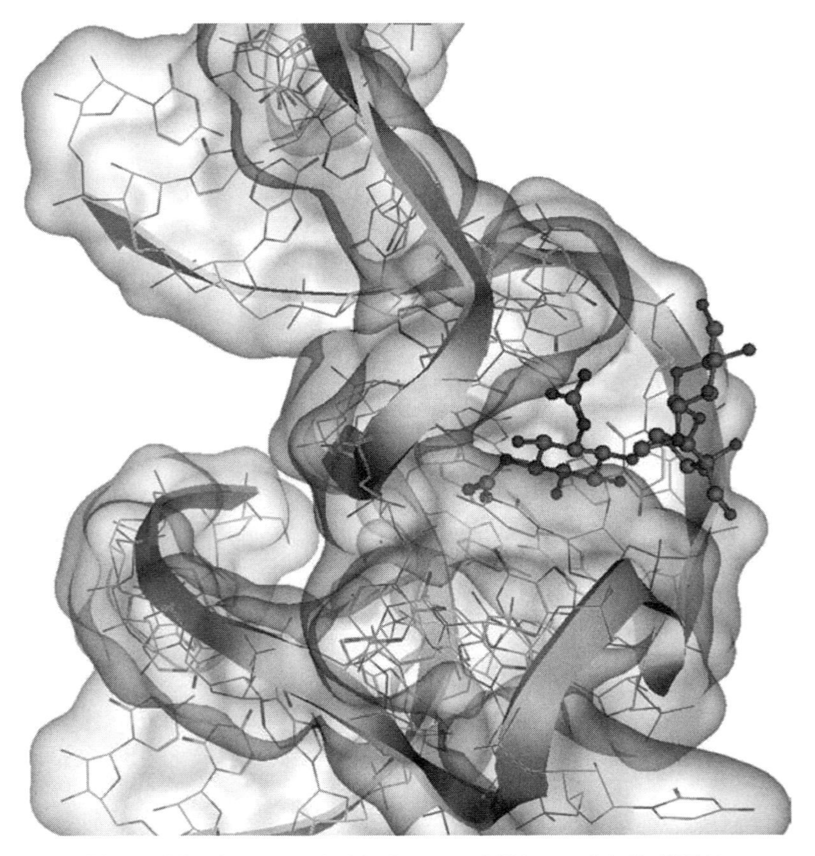

**Figure 7.3** Streptomycin binding to an RNA strand (pdb 1NTA).

Sometimes, docking programs will have additional options for localizing where the inhibitor can be located. Sometimes, it is useful to use molecular dynamics simulations to explore binding conformations.

As with the design of DNA-binding molecules, one side of the inhibitor is often exposed to the solvent, namely, the cytoplasm of the cell. The functional groups on this side of the molecule can be altered to alter pharmacokinetics and toxicity. Pharmacokinetic turning will be important in order to ensure that the drug can diffuse through the cell membrane.

In recent years, antisense compounds have received quite a bit of publicity. It is relatively easy to identify a short RNA sequence that can be bound by an antisense compound to downregulate protein production with very good specificity. The problems in bringing antisense drugs to market have been associated with the pharmacokinetics of making such compounds orally bioavailable, and capable of being transported into the cell and to the target RNA. At present, only one antisense drug has been approved by the FDA for prescription use: fomivirsen (trade name Vitravene, from Isis Pharmaceuticals), for the treatment of cytomegalovirus retinitis.

## 7.3 ALLOSTERIC SITES

Allosteric drugs bind to a secondary site on the protein. Thus, they can be designed to bind in the same way that a competitive inhibitor is designed—but that is not enough. Binding in the active site is not necessarily sufficient for an allosteric inhibitor to have activity. Thus, simple docking experiments are not enough to predict whether the drug will have activity. The way in which the allosteric drug interacts must be taken into account. Some allosteric compounds alter the shape or vibrational motion of the target, while others may donate or accept electrons or protons. This gives a secondary requirement for obtaining drug activity, which is just as important as the ability to bind at the correct site.

In spite of the additional design issues, allosteric drugs are very important. Allosteric binding often acts as a feedback mechanism to control a metabolic pathway. Because of this, an allosteric site may be the most effective target for a drug. Some drugs that were developed without knowledge of the target were later found to be allosteric regulators.

Another reason that allosteric binding is important is that it has the potential to upregulate a metabolic pathway. In the vast majority of cases, competitive inhibitors can only downregulate protein activity.

An example of an allosteric drug is paclitaxel, which binds to tubulin. Molecular dynamics simulations are necessary to understand how paclitaxel serves to act as an antitumor drug at a molecular level. Figure 7.4 shows another example of an allosteric compound, in this case binding to the caspase-7 protease.

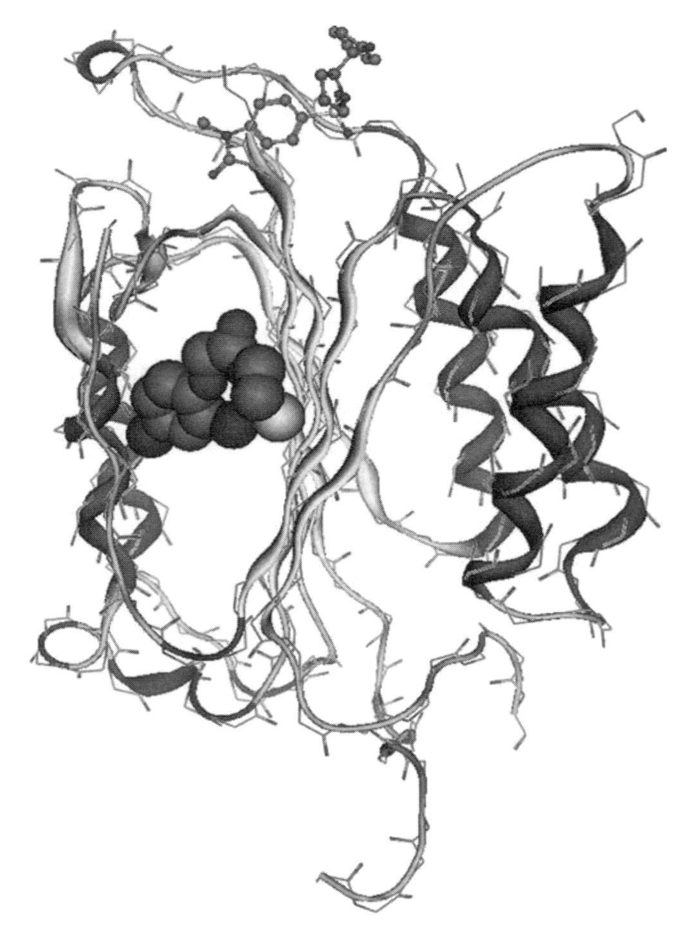

**Figure 7.4** Caspase-7 with a molecule in the active site shown as a ball and stick structure and a molecule in the allosteric site shown as a space-filling model (pdb 1SHJ).

## 7.4 RECEPTOR TARGETS

Some successful drugs bind to cell surface receptors, most notably G-protein-coupled receptors (GPCR). One advantage of this tactic is that the drug does not need to reach the cell interior, thus alleviating some bioavailability issues. Also, the drug can be a receptor agonist to initiate the receptor response, or an antagonist to shut down the receptor response.

Unfortunately, there are additional technical problems associated with the design of receptor-binding drugs. Since both agonists and antagonists bind to the receptor, simply designing a drug to bind, as is done for competitive inhibitors, is not sufficient to ensure the correct response. Most receptors are activated when a bound ligand induces a particular conformational change. Because of this, the way in which a compound binds, as well as its size, is crucial to obtaining the desired response. Thus, seemingly innocuous changes in the structure of a ligand can result in radical changes to its activity.

Because the induction of a new receptor conformation is the mechanism of action, the scoring functions included with nearly all docking programs tend to do a poor job at predicting compound activity against receptor sites. Some researchers do not use docking at all for receptor or allosteric targets. Others define a new scoring criterion, for example based on the distance between two atoms in residues on opposite sides of the binding site. As this implies, a docking tool that allows active site flexibility is required.

Pharmacophore and QSAR models for receptors may or may not appear reasonable without producing any useful results. There is no way to predict whether an automated pharmacophore or QSAR model-generating tool will work well for a receptor. The same program may work well for one receptor and not for another. However, a careful analysis of the model and how it compares with the geometry of the active site can give a good indication of whether it has identified features that correspond to the desired drug activity.

## 7.5 STEROIDS

Steroids fit into receptor sites, as shown in Fig. 7.5. Thus, at first glance, steroids appear to be capable of being modeled with a typical docking study. However, there are several subtle aspects of steroid modeling that must be taken into account.

The first item to note is that steroids are very lipophilic. Because of this, binding to the receptor will not be dominated by hydrogen bonding and charged group interactions, as is the case for many protein inhibitors. Steroid binding must be modeled with a force field that gives a good description of van der Waals interactions, $\pi$-system stacking, and, most importantly, subtle partial charge effects. Thus, it is important to validate the selection of a charge computation scheme for modeling steroid interactions.

Even more importantly, steroid receptors undergo a conformational change after ligand binding. This change is more complicated than a typical induced fit interaction. The complexity lies in the fact that the same receptor may undergo two different conformational changes, depending upon which steroid binds to it. Thus, modeling efforts must include a test for which conformational change is induced, in order to ensure that the desired conformational change is occurring.

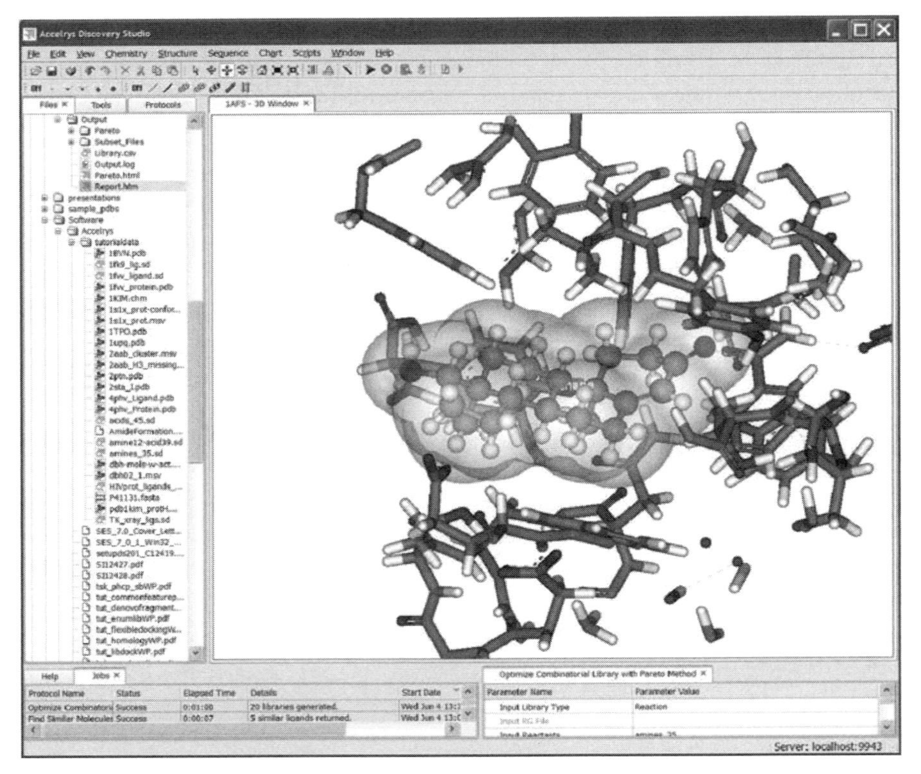


**Figure 7.5** Testosterone complexed to a dehydrogenase (pdb 1AFS). This example shows Discovery Studio, a full-featured drug design software package sold by Accelrys.

## 7.6 TARGETS INSIDE CELLS

The design of drugs for intracellular drug targets requires additional attention to pharmacokinetics. There are only a few ways for cells to get through a cell membrane. One is to be sufficiently lipophilic to pass into the membrane, and then out of the other side. Another is to be recognized by an active transport mechanism, which will thus transport the drug into the cell. When other approaches fail, a prodrug approach can be taken.

There are no direct measurements of how well compounds will traverse the cell membrane. In lieu of such measurements, researchers will use Caco-2 assays, or computational models designed to predict lipophilicity and Caco-2 results. It is understood that Caco-2 is not a direct measurement of bioavailability to the cell cytoplasm, but the Caco-2 results are generally considered to be proportional to cytoplasm bioavailability. Caco-2 assays are discussed further in Chapters 2 and 19.

Lipophilicity can theoretically be a more direct approach, although there is still some work involved to determine the optimal lipophilicity range for a given cell line and tissue. Lipophilicity is correlated with molecular weight, number of hydrogen bond donors/acceptors, and polar surface area. A further description of these prediction techniques is given in Chapter 19.

## 7.7 TARGETS WITHIN THE CENTRAL NERVOUS SYSTEM

The central nervous system (CNS) is surrounded by the blood–brain barrier (BBB). The BBB acts to prevent many of the substances found elsewhere in the body from getting to the sensitive organs in the CNS. The tight junctions between BBB cells, which are modified epithelial cells, allow only a miniscule percentage of water-soluble molecules to pass, making water-soluble drugs ineffective for CNS treatment. Most drugs do not cross the BBB, and particular attention must be paid in order to design drugs that do.

There are only a few strategies for designing drugs to cross the BBB. One option is to tune the lipophilicity of the compound to improve its passive diffusion. Indeed, the majority of CNS-active drugs have a log $P$ value of about 2. Likewise, avoiding compounds with log $P$ near 2 will prevent non-CNS drugs from producing CNS-related side effects. In general, CNS-active drugs are of lower molecular weight than other drugs, and have fewer polar functional groups. Because of the sensitivity to lipophilicity, even a single hydroxyl group added or removed can significantly change BBB penetration. Uptake by the CNS can also be improved by formulating the drug for intranasal administration, thus allowing it to enter via the olfactory lobe.

Another option is to make the drug, or its prodrug form, bind reversibly to one of the membrane-bound proteins. For example, prodrugs can be created by attaching the active drug to a monoclonal antibody (MAb) of the transferrin receptor or human insulin receptor. Of these two, there is experimental evidence that the insulin receptor is approximately ninefold more effective as a transport vector.

In rare cases, it may be possible to make the drug look very much like a molecule for which an active transport pathway is available. For example, drugs have been designed to bind to the transport proteins for glucose, amino acids, and nucleosides. At present, work is in progress to utilize the insulin receptor for drug transport (as opposed to the prodrug transport mentioned in the previous paragraph), but no drugs utilizing this mechanism have yet been approved. As a last resort, drugs that are administered in a hospital setting can be injected directly into the cerebrospinal fluid surrounding the spinal cord (intrathecal injection).

A few software packages are available that will predict a compound's BBB permeability. They predict the permeability based on an assumption that simple diffusion is the only mechanism by which the compound can access the CNS.

## 7.8 IRREVERSIBLY BINDING INHIBITORS

Inhibitors that bind irreversibly to the active site are sometimes called suicide inhibitors. These inhibitors are used once, and do not return to the bloodstream after finding their target. Also, the target protein is permanently deactivated, not just downregulated.

This scheme has both advantages and disadvantages. The advantage is that suicide inhibitors can be very potent. One disadvantage is that side effects can be severe if the drug binds irreversibly to an unintended target. Another disadvantage is that completely deactivating the protein target can be too severe an action and thus create severe side effects because of being too potent.

There are two ways in which these advantages and disadvantages are generally addressed. First, suicide inhibitors are most often intentionally designed to inhibit proteins in pathogenic organisms, where a high potency is not a problem. Second, specificity for just the intended target is a more stringent requirement for the design of suicide inhibitors than for the design of other drugs.

Suicide inhibitors typically utilize the enzyme's native reaction mechanism. For example, a suicide inhibitor for a protease may undergo the first step in the reaction in which a bond is formed between the inhibitor and the enzyme residue. However, if the inhibitor does not have a suitable leaving group, the subsequent step of cleaving the substrate and ejecting the pieces from the enzyme cannot occur. Examples of suicide inhibitors are penicillin, zidovudine (AZT), allopurinol, sulbactam, and eflornithine.

## 7.9 UPREGULATING TARGET ACTIVITY

It has already been mentioned that most drugs are designed to work through competitive inhibition. Competitive inhibition downregulates target activity. Upregulating target activity is more difficult, and not always possible.

The primary strategy in finding a way to upregulate target activity is to examine metabolic pathways carefully. Most metabolic pathways have feedback mechanisms to control their activity. It is sometimes possible to downregulate a protein at a different point in the metabolic pathway, in order to provide the feedback that will upregulate the action of the desired target. It is also possible to mimic the feedback signal by making a drug that will bind at an allosteric control site.

Lastly, the body's own signaling mechanism can be utilized. One example of this tactic is the creation of new steroidal compounds. Another is making a receptor agonist to initiate the receptor response, or an antagonist to shut down the response. Design of receptor agonists and antagonists can be good way to either upregulate or downregulate a metabolic pathway. An added advantage of this tactic is that it is not necessary for the drug to reach the cell interior.

With the existence of all of these atypical drug design scenarios, it sometimes seems as though there are more atypical cases than typical ones. Indeed, each drug design project must be approached as a unique case, rather than relying on a "cookie cutter" design process.

## BIBLIOGRAPHY

### Types of Drug Targets

Filmore D. It's a GPCR world. Mod Drug Discov 2004; 11: 24–28.

Schneider G, Baringhaus KH. Molecular Design. Weinheim: Wiley-VCH; 2008.

### Drugs Binding to DNA

Raber J, Zhu C, Eriksson LA. Theoretical study of cisplatin binding to DNA: The importance of initial complex stabilization. J Phys Chem B 2005; 109: 11006–11015.

Sadowitz PD, Hubbard BA, Dabrowiak JC, Goodisman J, Tacka KA, Aktas MK, Cunningham MJ, Dubowy RL, Souid AK. Kinetics of cisplatin binding to cellular DNA and modulations by thiol-blocking agents and thiol drugs. Drug Metab Dispos 2002; 30: 183–190.

Shaikh SA, Jayaram B. DNA Drug Interaction. Available at http://www.scfbio-iitd.res.in/doc/preddicta.pdf. Accessed 2008 Feb 26.

Tabernero L, Bella J, Alemán C. Hydrogen bond geometry in DNA-minor groove binding drug complexes. Nucleic Acids Res 1996; 24: 3458–3466.

### Drugs Binding to RNA

Bailly C, Colson P, Houssier C, Hamy F. The binding mode of drugs to the TAR RNA of HIV-1 studied by electric linear dichroism. Nucleic Acids Res 1996; 24: 1460–1464.

Wallace ST, Schroeder R. *In vitro* selection and characterization of streptomycin-binding RNAs: Recognition discrimination between antibiotics. RNA 1998; 4: 112–123.

### Allosteric Drugs

Madsena BW, Yeo GF. Markov modeling of allosteric drug effects on ion channels, with particular reference to neuronal nicotinic acetylcholine receptors. Arch Biochem Biophys 2000; 373: 429–434.

Xiao H, Verdier-Pinard P, Fernandez-Fuentes N, Burd B, Angeletti R, Fiser A, Horwitz SB, Orr GA. Insights into the mechanism of microtubule stabilization by Taxol. Proc Natl Acad Sci USA 2006; 103: 10166–10173.

## Steroid Binding

D'Ursi P, Salvi E, Fossa P, Milanesi L, Rovida E. Modelling the interaction of steroid receptors with endocrine disrupting chemicals. BMC Bioinformatics 2005; 6(4): S10–S17.

Wolohan P, Reichert DE. CoMSIA and docking study of rhenium based estrogen receptor ligand analogs. Steroids 2007; 72: 247–260.

## Designing Drugs for CNS Disorders

Pardridge WM. CNS drug design based on principles of blood–brain barrier transport. J Neurochem 1998; 70: 1781–1792.

## Designing Irreversibly Binding Inhibitors

Coombs GH, Croft SL, Molecular Basis of Drug Design and Resistance. Cambridge: Cambridge University Press; 1998.

Svendsen A, editor. Enzyme Functionality: Design, Engineering, and Screening. Boca Raton, FL: CRC Press; 2003.

Additional references are listed on the accompanying CD.

# 8

# COMPOUND LIBRARY DESIGN

Compound library design usually refers to the generation of a list of structures to be synthesized through combinatorial synthesis. There are a number of approaches that drug designers can utilize in performing this task. Designers may get fairly deep into the experimental plate design, or they may design a library as a collection of compounds that researchers would like to test, without regard for the synthesis route. There are a number of software tools for aiding with library design, which have a correspondingly diverse range of functionality.

## 8.1 TARGETED LIBRARIES VERSUS DIVERSE LIBRARIES

One issue to be considered is whether the library is to be narrowly focused (a targeted library) or very diverse. In the earlier stages of a design project, diverse libraries will often be used in order to explore a wide range of chemistries. Later, strongly focused groups of compounds (possibly differing only by a single functional group) will be synthesized and tested.

In general, it is easier to design a narrowly targeted library. This is done to explore possible derivatives of a known structure, usually for the purpose of increasing activity. It is typically done by first selecting a backbone structure (often a fused ring system), and then selecting synthons to be used to create derivatives of that structure. The designer can identify a point on

*Computational Drug Design*. By David C. Young

the molecule to be altered, and select a bioisosteric group of functional groups to put at that point. The term "synthon" refers to a functional group to be added at a particular point. Synthons are typically molecular fragments with unfilled valence positions, not reagents to be used in a synthetic reaction. Bioisosteric replacements are discussed in Chapter 5. Focused libraries are also easier to synthesize. It is often possible to use the same chemical reaction for all of the compounds, with just one reagent substituted.

Designing diverse chemical libraries tends to be a more difficult task. Researchers must contend with some rather difficult questions of chemistry:

- How diverse should the library be?
- Are there any two compounds that are too similar?
- Are there gaps in the chemistries represented, where an additional compound with those specifications should be included?
- Does the library span the space of known chemistries, or known drug-like chemistries?

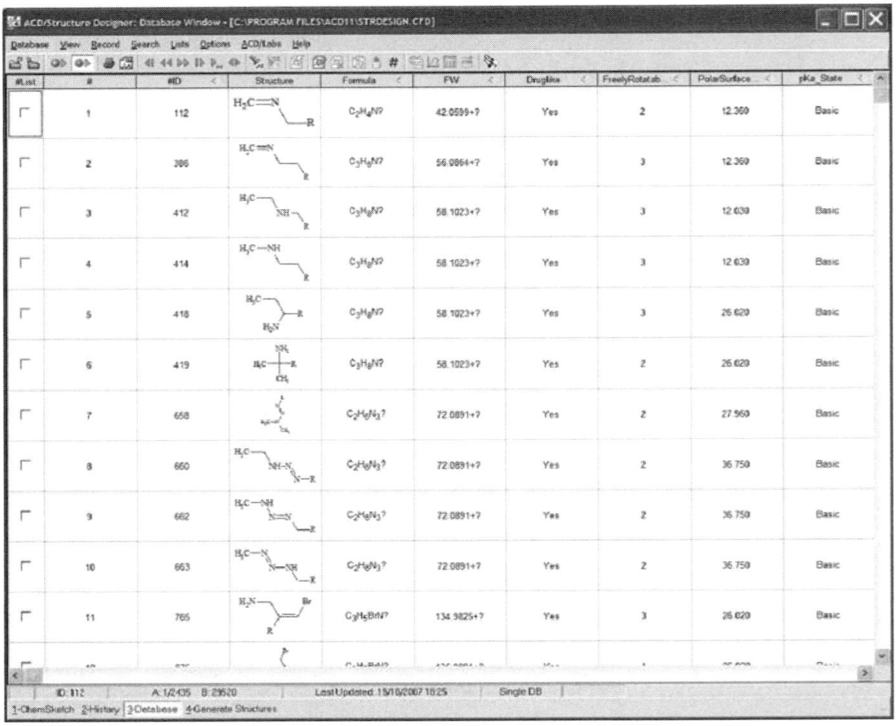

**Figure 8.1** The ACD/Structure Designer program makes modifications specifically for the purpose of improving the pharmacokinetic properties of an active lead.

Often, a diverse library is created by selecting compounds from a list of compounds already synthesized and available either in inventory or from commercial sources. This is more cost-effective than trying to synthesize a very diverse set of compounds for each drug design project.

Late in the drug design process, focused libraries are made for the purpose of improving bioavailability, half-life in the bloodstream, and toxicity. Figure 8.1 shows an example of a library design tool specifically for this purpose.

## 8.2 FROM FRAGMENTS VERSUS FROM REACTIONS

Most library design tools work by allowing the user to suggest the functional group lists; the tool can then generate structures for the compounds that would be created from those lists. When examining the "nuts and bolts" of how library design tools work, there are two different approaches for defining the reagent lists: the fragment approach and the reaction approach.

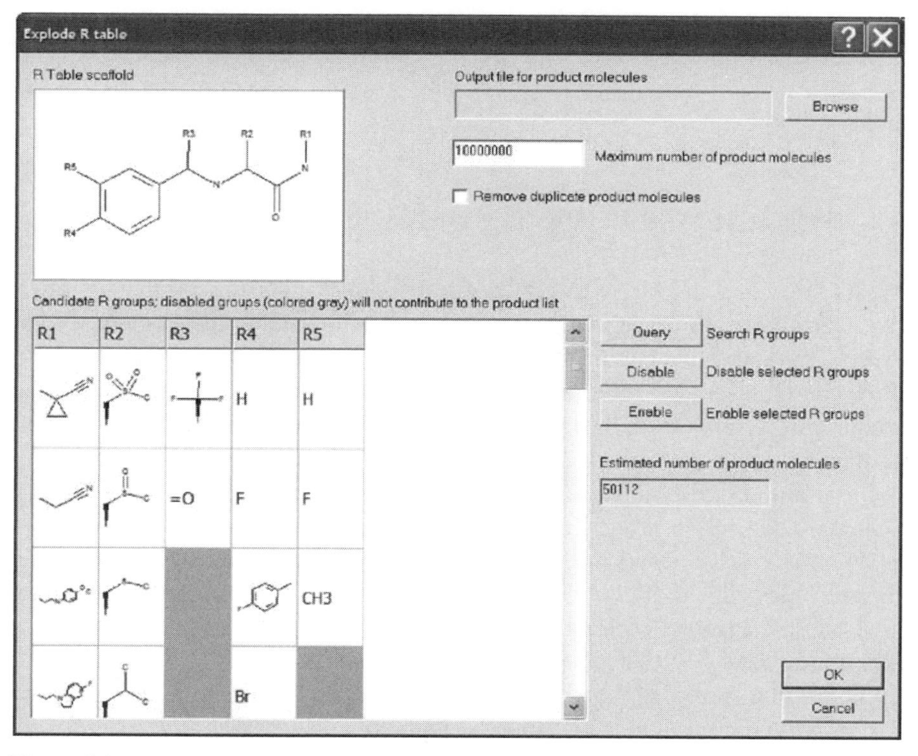


**Figure 8.2** Class Pharmer from Simulations Plus is a library design tool that uses a fragment-based approach.

In the fragment approach, backbones and side chains are defined with an open valence location defined as a dummy atom, instead of a hydrogen. This fragment is called a synthon. The program can then generate a list of product compounds by connecting the functional groups in the list to the backbones, knowing that the dummy atom is the point of connection. The advantages of this approach are that the researcher can stay focused on the resulting set of molecules and that there are no ambiguities in how the researcher intends to connect the pieces together. The disadvantage is that the design process is rather disconnected from the synthesis process. Some programs can generate fragment lists automatically from a list of compounds, and some require that every fragment be edited by hand to define the connection points. Figure 8.2 shows an example of a fragment-based library design program.

The alternative is to work with a piece of software that allows the researcher to define a chemical reaction and the list of reagents. The advantage of this is that it is closer to the synthesis route, so the results do not often come up with compounds that cannot practically be synthesized with the intended reaction. This can bring to light ambiguities in the synthesis when there are multiple functional groups, thus making it possible to create several different products from the chosen reactants.

Regardless of which type of library creation tool is used, the majority of the work at the library design stage is in the process of selecting the backbones and synthons.

## 8.3 NON-ENUMERATIVE TECHNIQUES

Most library design tools are enumerative techniques. This means that the functional group lists are used to generate structures, in the computer's memory, for every compound that can be synthesized in the library. The entire list of structures can then be feed into various types of prediction software. Enumerative techniques work well for designing small to moderate-size compound libraries, containing up to thousands or tens of thousands of compounds.

Non-enumerative techniques are useful for manipulating very large library designs. The backbone and functional group lists are generated, just as they are in an enumerative algorithm. However, in a non-enumerative algorithm, the structures for each individual compound are never generated. Group additivity methods are used to compute properties for the library, such as minimum, maximum, and estimated average molecular properties. However, only numbers that describe the entire library are generated in this way—values for individual compounds are never computed.

The advantage of non-enumerative methods is that they can be used to explore large, diverse chemical spaces representing billions of compounds, in a way that would not be practical, or even possible, using other techniques. This is a different type of software to work with. Users must get used to thinking in terms of a large theoretical chemical space, instead of being able to see individual molecular structures.

## 8.4 DRUG-LIKENESS AND SYNTHETIC ACCESSIBILITY

Chapter 2 discussed drug-likeness metrics. These metrics are one of the criteria used when selecting compounds from available sources to include in an initial diverse assay. Researchers also utilize their understanding of drug likeness when designing diverse libraries to be synthesized. The existence of the two different uses is why some drug-likeness metrics are easily understood rules, while others are more accurate, but obtuse, mathematical formulas. The structural rules are more easily applied to design work, and the mathematical predictions can be used to virtually test a library that has been designed *in silico*.

Synthetic accessibility can be a concern in designing compounds to be synthesized in the early stages of drug design efforts. At this point, some cost-effectiveness can be achieved by avoiding compounds that could take weeks or months to synthesize. In the late stages in the design process, sets of complex molecules with very subtle variations in functional groups will be created, and synthetic accessibility is for the most part ignored in favor of achieving the goal of creating a marketable drug.

The earliest and simplest metrics of synthetic accessibility used molecular weight, based on the simplistic ideal that larger molecules are more complex to synthesize. This has been supplanted by two different breeds of synthetic accessibility estimations, which give much more reasonable results.

One class of software for synthetic accessibility estimation comprises programs for the prediction of synthesis routes. These packages come up with a proposed synthesis scheme, starting from commercially available chemicals. One trivial result from this analysis is the number of steps in the synthesis scheme, which is a good way to estimate how difficult the synthesis will be.

A more complex analysis is performed by CAESA from SimBioSys Inc. CAESA is a web-based synthesis route program that provides a synthetic accessibility metric, along with an analysis of how that value was obtained, as shown in Fig. 8.3. The analysis done by CAESA includes the number of synthetic steps, availability of starting materials, and relative difficulty of synthesis of specific structural features. The disadvantage of this type of analysis is that it can be too time-consuming to run predictions for an entire database of millions of compounds.

The alternative is to use a cheminformatics metric designed for ranking compounds based on synthetic accessibility. A number of such metrics have

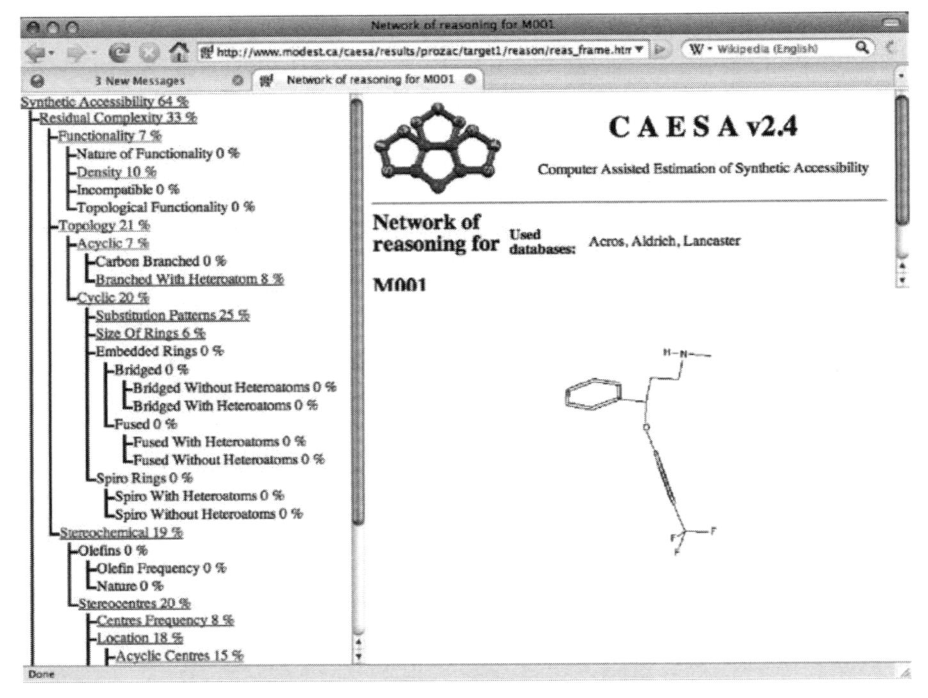


**Figure 8.3** CAESA from SimBioSys is a web-based tool for suggesting synthesis routes and giving a synthetic accessibility metric.

been proposed. Most look at points of molecular complexity, such as the number of rings, size of rings, number of heteroatoms, and number of chiral centers. There are also topological descriptors that examine the complexity of the loop-and-bond branch pattern in a mathematical way. Some utilize electronegativity to identify polarized bonds, which tend to be reactive functional groups. The advantage of these methods is that they can run predictions on millions of compounds in a reasonable amount of time. However, a very complex compound may be very easy to synthesize if it is only one functional group different from a commercially available chemical—but complexity metrics would still predict that it is very difficult to synthesize. Moreover, the more complex compounds may be the ones that make the best drugs.

One of the simplest complexity metrics, introduced by Whitlock, is based on bonds and rings. In this metric, the complexity formula consists of adding 4 times the number of rings, 2 times the number of nonaromatic unsaturations, 2 times the number of chiral centers, plus the number of heteroatoms. The advantage of this method is that it is easily understood and easily programmed into software for mass analysis of chemical structures. The disadvantage is that a compound with several single ring substituents can be rated as having the same complexity as a compound with a complex fused ring system, even though the fused heterocyclic systems are typically more complex to synthesize.

Barone and Chanon put together a metric designed to keep the simplicity of Whitlock's method, while improving the ranking of ring systems. This method is based on atoms, rather than bonds. For the atom-type contribution, the ranking is based on the number of hydrogen atoms. It includes the sum of the number of $XH_3$ groups, 2 times the number of $XH_2$ groups, 4 times the number of XH groups, and 8 times the number of carbon atoms without hydrogen attached. These atom contributions are multiplied by 3 if X is a carbon, and multiplied by 6 for any heteroatom. The ring contribution term consists of 6 times the number of atoms in the ring. This results in giving fused ring systems a higher weight, because atoms appearing in more than one ring are counted multiple times.

Bertz took a graph-theoretical approach to creating a complexity metric. He defines a total complexity as the sum of a complexity due to connectivity and a complexity due to heteroatoms. These are computed using an information content formula, which sounds sophisticated but is really just a logarithmic function.

A metric based more directly on the synthesis process itself involves counting the number of "strategic bonds" identified by a synthesis route prediction program, such as LHASA. This has been shown to rank compounds about the same as some of the graph-theoretical methods.

## 8.5 ANALYZING CHEMICAL DIVERSITY AND SPANNING KNOWN CHEMISTRIES

The issue of screening diverse libraries of chemical compounds was mentioned earlier in this chapter. This brings up the question of how chemical diversity is defined, and how to design or analyze a library for diversity. A number of techniques and software packages have been developed to address this issue. These can be used to answer the following questions:

- Does a compound library contain gaps in the represented chemistries?
- Does library A contain compounds that fill gaps in the diversity of library B?
- How can a group of *n* compounds be selected from a database of available compounds such that the group has the maximum amount of chemical diversity?
- Does a compound library span the space of known chemistries or just of drug-like chemistries?

Note that there are several questions that cannot be answered by these techniques. First, a space of known chemistries or drug-like chemistries can be defined by profiling a large database of compound structures, but this does not tell us anything about the space of all possible chemistries. Second, it is not possible to have the program design a compound to fill a void in the chemical space. In fact, filling such voids is a rather difficult task even for an experienced chemist, owing to the fact that the definition of such chemical spaces becomes somewhat mathematically obtuse, thus making it difficult to back-engineer what type of molecule could fill a given niche.

In order to define a chemical space, the researcher or software developer must choose some type of mathematical description of the molecule. The numbers describing the molecule may be used directly as the values along axes in the space, or they may be used more indirectly by defining axes derived from these numbers, such as via a principal components analysis. The following are some of the types of mathematical descriptions that have been utilized for this purpose.

***8.5.1.1 1D Molecular Fingerprints*** One-dimensional fingerprints are strings of binary bits, usually thousands of bits long. Each bit represents a yes/no result as to whether or not the molecule contains the chemical feature represented by that bit. The assignment of structural motifs to bits is often wrapped, thus using each bit for indicating one of several different structural motifs. This gives bit strings of a more manageable size, and avoids having every string come out mostly zeros. This wrapping algorithm has the potential to introduce occasional errors, but does so extremely rarely, so the benefits outweigh the liabilities.

Fingerprints can be compared with a distance metric, such as the Tanimoto distance, in order to indicate how similar the compounds are. For reasonably similar compounds, the distances between 1D fingerprints usually looks reasonable to a chemist's intuitive feeling for similarity. However, when compounds are rather dissimilar, the 1D fingerprint distances tend not to agree with a chemist's opinion. Chemists looking at these results tend to agree that compounds shown to be dissimilar by the fingerprints *are* dissimilar, but a chemist may not necessarily agree with the ordering of which compound is the least similar to a given structure. Thus, 1D fingerprints can be used to select sets of compounds that are diverse by some metric, but not necessarily what an individual scientist would have selected by hand.

***8.5.1.2 2D Molecular Fingerprints*** Two-dimensional fingerprints are matrices, which may represent connectivity, polarizability, molecular weight, charge, or hydrogen bonding. As such, 2D fingerprints can be a somewhat more accurate description of the molecule than 1D fingerprints. In order to compute distances, 2D fingerprints may be compared through eigenvalues, principal components, or some other mathematical manipulation. This gives a good representation of structural diversity. However, it does not necessarily imply diversity in terms of how the molecule will interact with a protein's active site.

Some variations on the 2D fingerprint theme include fingerprints derived from pairs of atoms, or layers of atoms out from a given center. These give a more well-defined distance relationship, roughly analogous to a radial distribution function. Atom pairs can be determined by counting the number of bonds between atoms or using distances from the Cartesian coordinates to give a type of 3D structural key. The primary weakness of keys based on Cartesian coordinates is the difficulty in taking conformational flexibility into account. Although 3D techniques sound like a good idea, the existing ones have been shown to perform poorly in several comparison studies.

***8.5.1.3 Molecular Properties*** Properties such as molecular weight, log $P$, dipole moments, and surface area can be used to define a chemical space. The advantage of this is that some of these metrics correspond directly to pharmacokinetic properties such as passive intestinal absorption. The disadvantage is that there is less correlation to chemical structure. Multiple property values may be combined into one vector, and then the vector distances computed.

***8.5.1.4 Pharmacophore Descriptions*** Analysis of diversity can also be accomplished by using pharmacophore features. For example, three pharmacophore features can be chosen, and the distances between them will uniquely define a triangle. The advantage of this is that it probably has the closest correlation to the way in which molecules will actually interact with the target's active site. It is much more difficult to work with more than three features, since distances are nonunique (e.g., having the same set of distances describing two different enantiomers). One solution to this problem is to encode the pharmacophore information in a binary bit string to create a pharmacophore-based fingerprint.

***8.5.1.5 Ring Statistics*** Looking at the types of rings present in the molecule gives a reasonable account of the various fused ring backbones present in the space, as well as identifying systems that differ by a bulky ring-containing side chain. However, this method completely ignores the presence of many types of functional groups. It sometimes uses topological indices. Some topological indices reflect rings only, while others can reflect branching, saturation, and heteroatom content as well. Topological indices can be based on the adjacency matrix, topological distance matrix, molecule centroid, or information-theory analysis.

***8.5.1.6 Side Chain Fingerprints*** The use of side chain fingerprints gives emphasis to the groups interacting with the active site, at the expense of completely ignoring the nature of the backbone. This is generally a better approach for analysis of large focused libraries than for analysis of very diverse libraries.

Creating a mathematical measure of chemical diversity is an extremely difficult task. An even harder task is defining a space of all of the possible chemical structures. Nonetheless, the analysis of such a space would give researchers a way to determine if there were chemical motifs that are not represented at all in a compound library. Only a few methods for spanning all of chemical space have been proposed. One of the most widely used of these is the BCUTs method.

In BCUTs, each compound is located at a point in a six-dimensional space, determined by the lowest and highest eigenvalues of the charge matrix, polarizability matrix, and a hydrogen bond matrix. This six-dimensional space can be analyzed to determine if it contains any regions that are not occupied by any compounds. This is often done by binning the number of compounds in various regions of the space. Note that this space is the space of all chemistries, not just drug-like compounds.

## 8.6 COMPOUND SELECTION TECHNIQUES

On the first pass of a drug screening effort, the most efficient option is to assay compounds that are already on site or available for sale. The task then becomes one of selecting which of the millions of available compounds to test. When selecting a diverse library of compounds, several different selection criteria are usually applied simultaneously. The first is to select compounds that are reasonably drug-like. The second is to select a very diverse sample of compounds. The third is to avoid the selection of a large number of very similar compounds. It is often desirable to have two or three compounds from each group of similar compounds, in order to minimize the potential harmfulness of false-negative results in the initial screening. This last point is particularly salient, because it is often a shortcoming of many library selection software packages. The following paragraphs discuss some of the existing compound selection algorithms.

Clustering algorithms group together similar compounds. Once the compounds have been grouped together, representative compounds from each group can be chosen. Often, the centroid of the group is selected. There are cluster sampling algorithms that find essentially the same result by analyzing a nearest neighbors list, without actually clustering the compounds.

There are also hierarchical clustering algorithms that create a tree or dendrogram view of compound similarity, with more closely similar compounds diverging lower on the tree. Sampling from dendrograms can performed by selecting compounds from each group as they are grouped at one particular level of the dendrogram.

The maximin algorithm successively selects compounds by choosing the one that is the most unlike any of those selected thus far. This allows the user to define the number of compounds selected and obtain a near-optimal choice of the most diverse library possible for the desired number of compounds.

There are statistical sampling techniques based on experimental design techniques. However, there seems to be some contention over how well these algorithms work. It is possible that these techniques can inadvertently sample just selected regions of the space.

Some of the more computationally intensive algorithms create many sets of compounds, and then use some mathematical criteria to choose the best set. The most computationally intensive sampling algorithms look at many possible conformers of each molecule, along with their charge distribution, to select an optimal set. Such techniques are attractive for their direct mapping

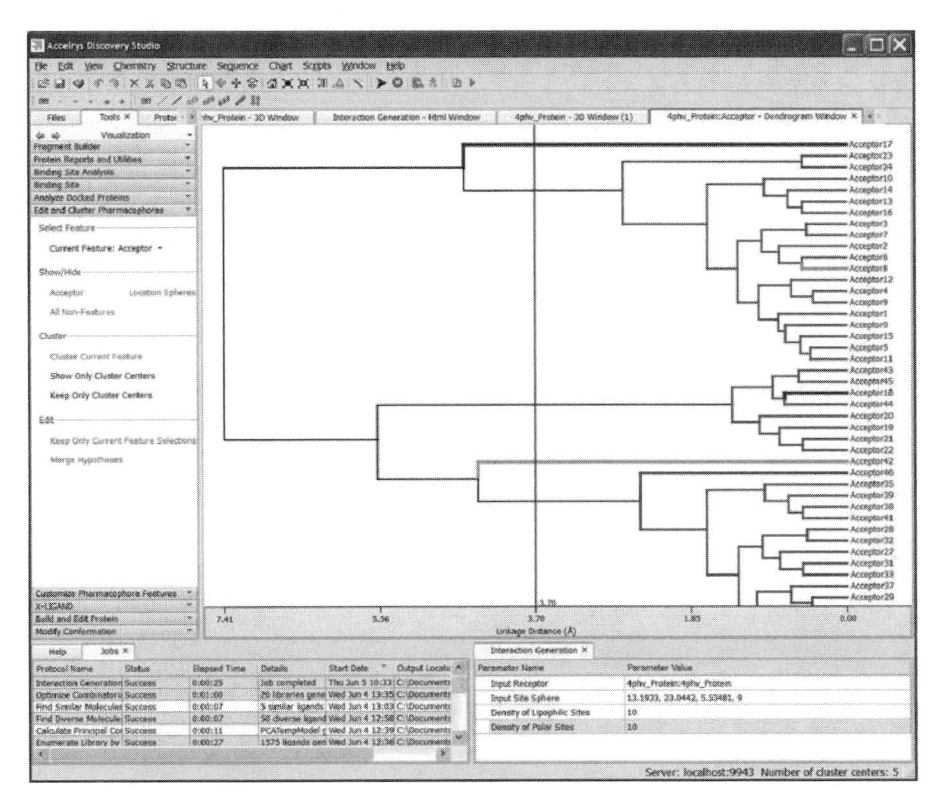


**Figure 8.4** A dendrogram view of a hierarchical clustering, in the Discovery Studio software from Accelrys.

onto a pharmacophore-type description. However, at present, these are probably limited to the analysis of a few thousand compounds. A less intensive variation on this theme is the analysis of vectors connecting selected functional groups, such as is done in the HookSpace method.

Most selection programs have some way to visually show the user what the space of compounds looks like, and what is being selected. For the hierarchical clustering methods, a dendrogram (tree) view is a natural visual representation, as shown in Fig. 8.4. A single number, such as molecular weight, can be visualized as a histogram, but using multiple histograms to visualize a half dozen properties becomes confusing. Two of the most useful of the visualization methods are Kohonen maps and nonlinear maps (NLM, also called SAR maps or SAR landscapes), both of which can map a multidimensional non-Euclidean dataset into points on a 2D picture. These 2D pictures are such that compounds that are similar to one another are mapped as points close to one another; the actual distance is not exactly proportional to the value of a similarity metric. Figure 8.5 shows a nonlinear map. The same types of

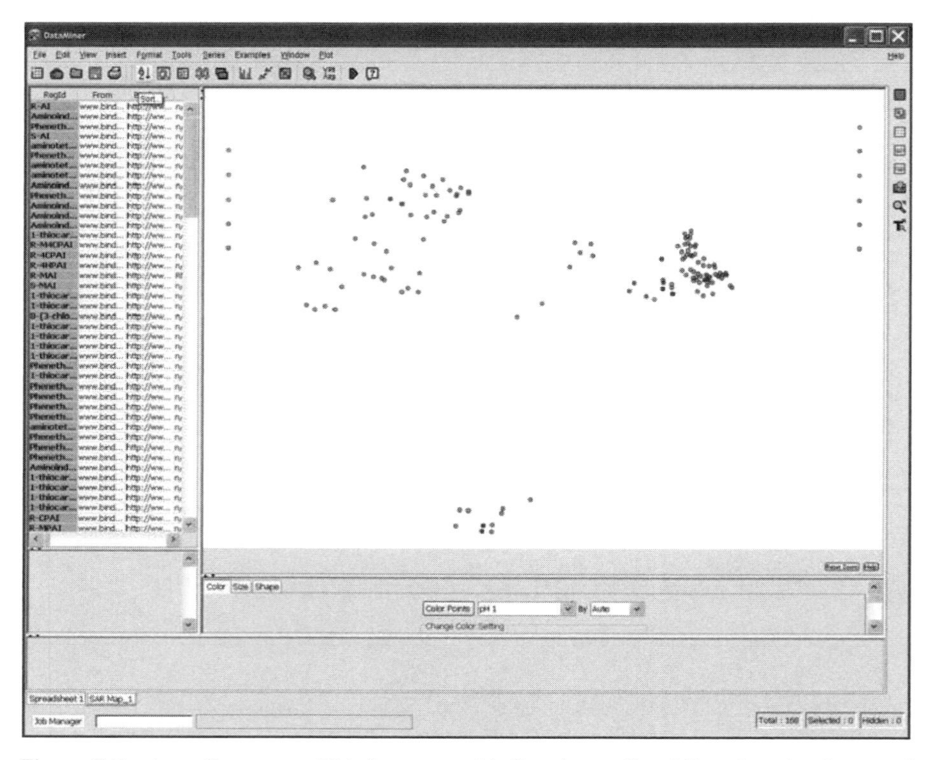


**Figure 8.5** A nonlinear map. This is generated in Benchware DataMiner (previously named SARNavigator) from Tripos Inc. The points around the periphery are singleton molecules that are not similar to any other compounds in the library.

tools are also used for analysis of screening results to identify clusters that correlate with compound activity.

Kohonen maps typically place compounds on the map based on a set of descriptors, as would be used for a QSAR or principal components analysis (PCA) prediction. The advantage of a Kohonen map is that there is a direct correlation with the drug having high or low efficacy. The disadvantage is that it is only as good as the ability of the descriptors to predict activity were they to be put into a QSAR or PCA equation—which are not generally the best methods for predicting activity.

Nonlinear maps cluster compounds near one another based on chemical structure. Their advantages are that they work even when there is not an accurate conventional QSAR equation, and that they can be used to identify false negatives and false positives. The disadvantage is that they do not give an indication of whether a compound will be active—only whether it is similar to other active compounds.

Compound libraries can be either selected from sources of available compounds or designed and created using combinatorial synthesis techniques. Early in the drug design process, diverse libraries of compounds are tested, whereas later in the process, the libraries become more focused. Diverse libraries are most often selected from available compounds. Focused libraries are most often designed and synthesized specifically for a given drug design project. A computational chemist should be familiar with both selection and design tools.

Libraries may be designed to be synthesized or by selecting compounds from lists of available structures. In either case, it is necessary to keep an eye on how the structures fit into the design project and the entire space of possible compounds. Ultimately, the best laid plans can be constrained by experimental limitations.

## BIBLIOGRAPHY

### Library Design

ABeavers MP, Chen X. Structure-based combinatorial library design: Methodologies and applications. J Mol Graph Model 2002; 20: 463–468.

English LB, editor. Combinatorial Library Methods and Protocols. Totowa, NJ: Humana Press; 2002.

Brown RD, Martin YC. Use of structure–activity data to compare structure-based clustering methods and descriptors for use in compound selection. J Chem Inf Comput Sci 1996; 36: 572–584.

Ghose AK, Viswanadhan VN, editors. Combinatorial Library Design and Evaluation; Principles, Software Tools and Applications in Drug Discovery. New York: Marcel Dekker; 2001.

Goodnow RA, Gillespie P, Bliecher K. Chemoinformatic tools for library design and the hit-to-lead process: A user's perspective. In: Oprea TI, editor. Chemoinformatics in Drug Discovery. Weinheim, Wiley-VCH; 2004. p. 381–435.

Jamois EA. Reagent-based and product-based computational approaches in library design. Curr Opin Chem Biol 2003; 7: 326–330.

Lowrie JF, Delisle RK, Hobbs DW, Diller DJ. The different strategies for designing GPCR and kinase targeted libraries. Comb Chem High Throughput Screen 2004; 7: 495–510.

## Synthetic Accessibility

Allu TK, Oprea TI. Rapid evaluation of synthetic and molecular complexity for *in silico* chemistry. J Chem Inf Model 2005; 45: 1237–1243.

Baber JC, Feher M. Predicting synthetic accessibility: Application in drug discovery and development. Mini Rev Med Chem 2004; 4: 681–691.

Barone R, Chanon M. A new and simple approach to chemical complexity. Application to the synthesis of natural products. J Chem Inf Comput Sci 2001; 41: 269–272.

Bertz SH. The first general index of molecular complexity. J Am Chem Soc 1981; 103: 3599–3601.

Rűcker C, Rűcker G, Bertz SH. Organic synthesis—art or science? J Chem Inf Comput Sci 2004; 44: 378–386.

Whitlock HW. On the structure of total synthesis of complex natural products. J Org Chem 1998; 63: 7982–7989.

## Diversity

Agrafiotis DK. Diversity of chemical libraries. In: Von Ragué Schleyer P et al., editors. Encyclopedia of Computational Chemistry. New York: Wiley; 1998. p. 742–761.

Agrafoitis DK, Myslik JC, Salemme FR. Advances in diversity profiling and combinatorial series design. Mol Divers 1998; 4: 1–22.

Bartlett PA, Entzeroth M. Exploiting Chemical Diversity for Drug Discovery. London: Royal Society of Chemistry; 2006.

Chaiken IM, Janda KD, editors. Molecular Diversity and Combinatorial Chemistry: Libraries and Drug Discovery. Washington: American Chemical Society; 1996.

Gordon EM, Kerwin JF, editors. Combinatorial Chemistry and Molecular Diversity in Drug Discovery. New York: Wiley; 1998.

Kitchen DB, Stahura FL, Bajorath J. Computational techniques for diversity analysis and compound classification. Mini Rev Med Chem 2004; 4: 1029–1039.

Lewis RA, Pickett SD, Clark DE. Computer-aided molecular diversity analysis and combinatorial library design. Rev Comput Chem 2000; 16: 1–51.

Martin EJ, Blaney JM, Siani MA, Spellmeyer DC, Wong AK, Moos WH. Measuring diversity: Experimental design of combinatorial libraries for drug discovery. J Med Chem 1995; 38: 1431–1436.

Pearlman DA, Smith DH. Novel software tools for chemical diversity. Perspect Drug Disc Des 1998; 9/10/11: 339–353.

Pirrung MC. Molecular Diversity and Combinatorial Chemistry. Amsterdam: Elsevier; 2004.

Additional references are contained on the accompanying CD.

# PART II

# COMPUTATIONAL TOOLS AND TECHNIQUES

# 9

# HOMOLOGY MODEL BUILDING

There will be times when a researcher wants to use structure-based drug design techniques, but is hampered by the lack of a crystal structure for the target protein. In this instance, the next best options are either building a homology model of the entire protein or using a 2D NMR structure of the active site region. Often, these two options are used together. A homology model gives a model of the entire protein, while a 2D NMR structure is based on limited data points. However, the 2D NMR analysis of the active site region can serve to verify that the homology model is reasonable. Homology modeling building is sometimes referred to as a knowledge-based technique, or as comparative protein modeling.

The idea behind homology modeling is fairly simple. A researcher has the primary sequence for a protein, and needs to find a three-dimensional structure, without waiting for crystallography results. A search is carried out on the primary sequence of proteins for which crystal structures are available. This search is set up to find regions that are identical, or nearly so, between the two proteins. Then the known crystallographic coordinates for an entire protein, or for various sections of known proteins, can be used to piece together a three-dimensional geometry for the unknown protein. This involves putting the pieces together, optimizing the geometry, and testing to see that a reasonable structure has been obtained.

*Computational Drug Design.* By David C. Young

## 9.1 HOW MUCH SIMILARITY IS ENOUGH?

Since the homology model building process is dependent upon utilizing crystal structure coordinates for similar proteins (called the template), a crucial factor to consider is how similar the unknown sequence should be to the template protein. A number of metrics have been suggested for this. One of the most conservative metrics suggests that there should be over 70% sequence identity with the template, in order to get a homology model that can be trusted. Other metrics suggest having over 30% or 40% sequence identity with the template. One study showed that having 60% or more sequence identity gave a success rate greater than 70%. Note that, even with high sequence identities, as many as 10% of homology models may have a root mean square deviation (RMSD) greater than 5 Å.

In order to clarify the seemingly disparate metrics mentioned in the previous paragraph, Rost performed a large study looking at how much sequence identity is needed to get a good homology model the majority of the time, as a function of the number of aligned residues. For a small sequence of 25 aligned residues, 60% identity was necessary. For a large region of 250 aligned residues, templates with over 20% identity could give good homology models. The metrics used by Rost are somewhat less conservative than some of the other metrics. Rost's results also reflect improvements in homology model software and methodology compared with earlier work. A number of software packages can display a graphical representation of a strong homology as shown in Fig. 9.1 or a poor homology as shown in Fig. 9.2.

The above paragraphs have described metrics based on percent identity for whether templates are good enough. Percent similarity is also a useful metric to examine. If several potential templates have essentially the same percent

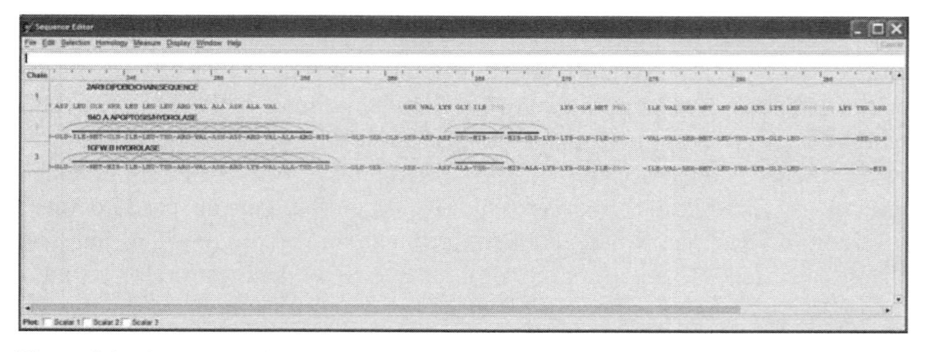

**Figure 9.1** Sequence alignment screen shot from the MOE software package (Chemical Computing Group). In this image, the residues are colored according to function, bars below the sequence indicate structural features, and arcs indicate hydrogen bonds. In this example, there is a strong homology between sequences for caspase-3 and caspase-7.

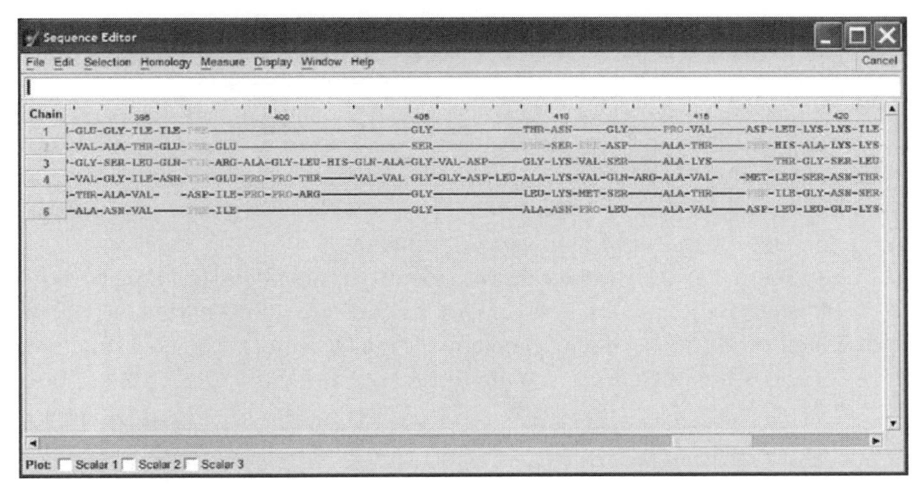


**Figure 9.2** Sequence alignment for sequences with a poor homology, generated with the MOE software.

identity, then the one with the highest percent similarity may be chosen. Researchers may also choose the one in which the crystal structure has the best resolution. When only structures for sequences with low similarity are available, threading- and fingerprint-based algorithms tend to give the best results.

## 9.2 STEPS FOR BUILDING A HOMOLOGY MODEL

Homology model building is a multistep process. The process often involves iterating back to previous steps, once a model has been created and validated. The point to which the iteration goes back will depend upon the specific problem identified in validation. Thus, the general procedure is as follows:

*Step 1*: Template identification
*Step 2*: Alignment between the unknown and the template
*Step 3*: Manual adjustments to the alignment
*Step 4*: Replace template side chains with model side chains
*Step 5*: Adjust model for insertions and deletions
*Step 6*: Optimization of the model
*Step 7*: Model validation
*Step 8*: If errors are found, iterate back to previous steps

The following sections discuss each of these steps.

### 9.2.1 Step 1: Template Identification

The object of template identification is to find the protein that has a known crystal structure and has a very high percent identity to the model sequence—in other words, one that is very close in terms of evolutionary distance. Ideally, the template should be one protein that has a high percent similarity to the entire unknown sequence, and large regions of similarity along the entire length of the unknown sequence.

The structure with the highest percent identity is not necessarily the best one to use. It is advisable to look at several of the best structures, particularly comparing their resolution ($R$-factor) and anisotropy ($B$-factor). The best structures have $B$-factors less than 15 Å$^2$, while those with $B$-factors greater than 50 Å$^2$ should be avoided if possible. It might be better to choose a template that is a slightly poorer match if the best match is a structure of marginal quality. When all of the options are of less than desired quality, hydropathy plots can be used to choose the one that should have the most similar fold.

Sometimes, it is not possible to find a single protein that forms a good template for the entire unknown sequence. The alternative is to find several proteins, each of which has a section with a high similarity to a section of the unknown. The entire structure can then be put together from these structurally conserved regions (SCR) and any missing variable regions (VR). Ideally, each region should show a high percent identity: 40% or greater. There are software packages that automate the searching for both of these options. The difference between this multiple-template process and the preferred single-template process is illustrated in Fig. 9.3.

At this step of the process, it is wise to save perhaps the dozen best potential template structures. The one that has the highest percent identity may not be the one that looks the best when examined in subsequent steps of the process.

There are times when there is just no good template available. There may be no crystal structures with high percent identity, or large regions with a high identity. Sometimes, the only options are a species with a very low percent identity or bits and pieces of a dozen residues or less. This happens more often when determining structures for membrane-bound proteins, which are more difficult to analyze crystallographically. In these cases, the researcher may simply have to admit that they cannot construct a homology model with any confidence in its accuracy. It is generally better to work around the lack of a target structure than to spend large sums of money on research based on an incorrect structure.

### 9.2.2 Step 2: Alignment between the Unknown and the Template

An alignment is performed to see how well the unknown sequence aligns with the template. If multiple templates seem promising, a multiple sequence alignment may be performed to examine all of them at once. It is common to find

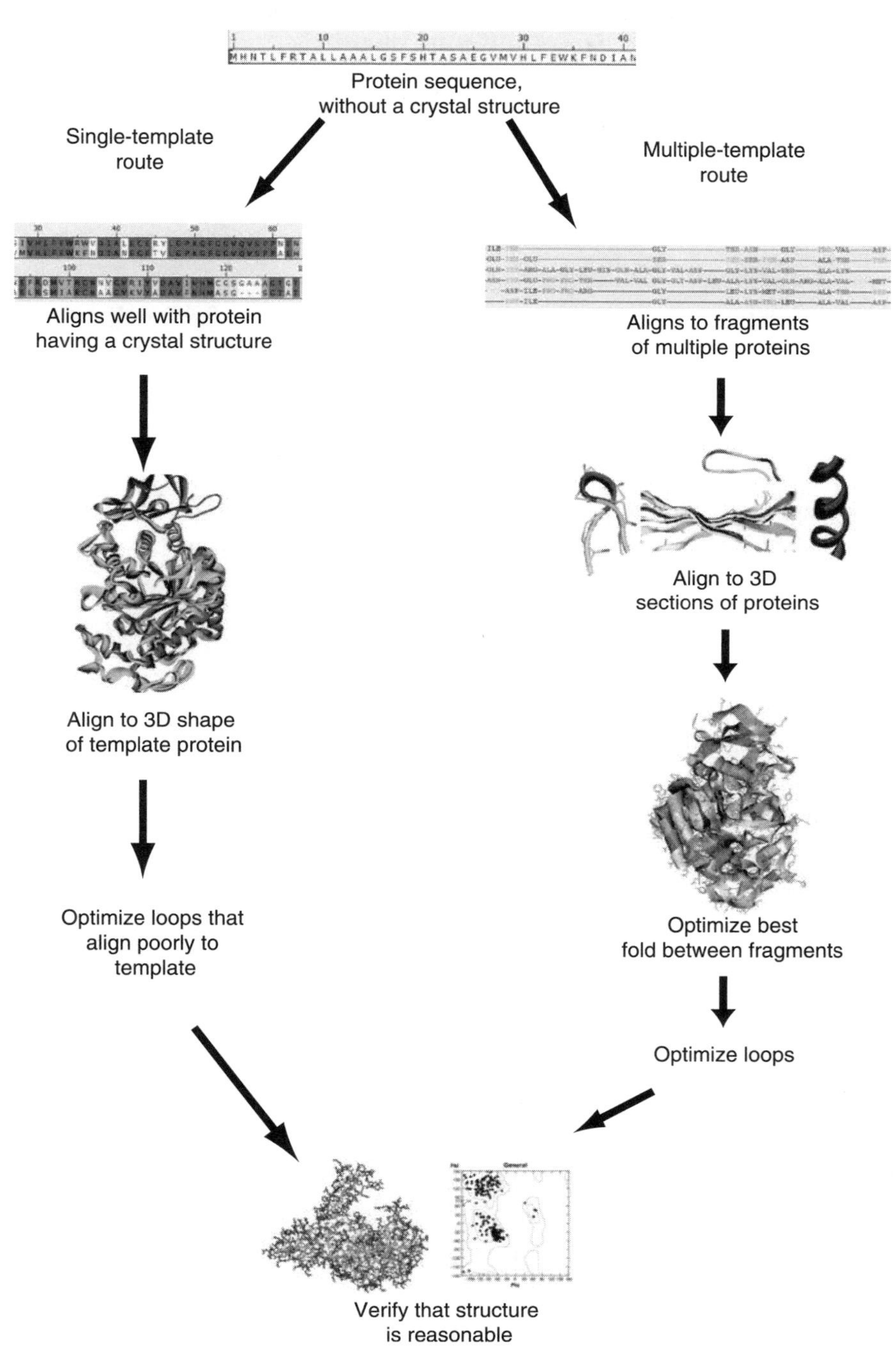


**Figure 9.3** Flow chart of the homology building process from a single template and from multiple templates.

spots where the two sequences differ by insertions or deletions, usually of about half a dozen residues. Total percent similarity aside, it is best to have a similar-length template with moderately large sections of identical sequence. The second best scenario is to find a template with large regions of similarity. It is least desirable to have many small regions of identity or similarity.

There are a number of good software tools for displaying alignments and multiple sequence alignments. Some of these tools can be used to carry out the manual adjustments to alignment discussed in the next subsection. Most such tools have options to color residues based on various criteria, as shown in Fig. 9.1.

### 9.2.3 Step 3: Manual Adjustments to the Alignment

There are a number of situations in which there is more than one way to assign a reasonable alignment between the unknown and the template. Consider the most common situation in which there is an insertion of a few residues in the template, relative to the unknown. In this situation, a reasonable alignment may be obtained by assigning a few unknown residues to be either at the end of one similarity region or at the beginning of another region. In this case, it is often best for the user to make a logical decision. In this particular option, there are going to be a few residues cut out of the template sequence in order to generate an unknown structure. Examining the three-dimensional geometry of the aligned species can determine whether the bond made to connect the two similar regions of the unknowns together will be a short bond or a long bond. Most often, the option that gives the shorter bond will give the better homology model.

A more difficult situation is when there is very marginal identity, with sections of only a couple of residues at a time being identical. There can even be situations where every other residue matches. There can also be cases where none are identical, but the region can be assigned if all of the similar residues can be aligned. In this case, a multiple-sequence alignment can give the required insight. For example, the alignment of sequence A to sequence B may be ambiguous, but sequence A may align to sequence C, which aligns to sequence B. The structure can then be generated either from pieces of each sequence or from a geometric average of multiple structures.

Note that the scenario in the previous paragraph is one of the least preferable situations. This is because it entails the highest probability of producing an incorrect homology model. It is generally preferable to take regions of strong similarity from multiple proteins, if such can be found. However, the technique described in the previous paragraph is recommended if no better template choices can be found.

It was mentioned above that the homology model may be built up piecewise by taking three-dimensional coordinates from several different crystal structures. If this is the case, the separate sections will be linked together after performing alignment adjustments. The three-dimensional positioning of these separate sections must be done carefully, as this can be the source of very large errors in the final model. Often, the checks discussed in Section 9.2.7 are run at this point to verify that a reasonable positioning has been obtained.

When there are no known sequences with good homology, even piecewise, a threading algorithm may be used. This algorithm tries multiple alternative folding motifs, and then checks whether the resulting structure is reasonable in order to determine which of the marginal fits gives the best end result. Some threading algorithms are very robust in comparing sequences with gaps and deletions. There are also sequence fingerprint techniques, which are a very different algorithm but are like threading in that they work well to align regions of marginal similarity.

### 9.2.4 Step 4: Replace Template Side Chains with Model Side Chains

For those residues that are not identical, the template side chains must be replaced with the appropriate side chain from the model compound. The only difficult part of this procedure is optimizing the three-dimensional position of the new side chain. Often, this is done by performing an optimization on a small conformation search on just the side chain being inserted, while holding the other side chains around it in a fixed position. Many of the best side chain placement algorithms are knowledge-based, meaning that they utilize information such as the most common conformational position for various residues.

At a later step in the process, the entire model will go through optimization and validation. If the validation indicates that side chain positions may be questionable, it may be necessary to go back and use some of the second best conformers from the original side chain replacement.

### 9.2.5 Step 5: Adjust Model for Insertions and Deletions

Insertions and deletions are often found in the loops that interconnect sheets and helices. These regions are often the source of the most uncertainty in atomic positions, both when building homology models and when examining X-ray crystallographic results. This is because they tend to be the floppiest parts of the protein, even in its folded form. They tend to be the regions where a loop folds over the active site, as well as the regions that are most susceptible to being affected by crystal packing forces. Thus, these regions often show some anisotropy even in the experimental crystal structure.

Once the correct residues are in place, their positions must be optimized. A simple optimization is a good first step. In the case where several residues are deleted from the template structure, an energy minimization may be sufficient to obtain a reasonable structure. When residues are inserted, relative to the template structure, it is necessary to perform some form of conformation search to find the best conformer for the loop. There are a number of good algorithms for doing this conformation search on just the loop, including, among others, molecular dynamics, grid searchers, tabu searches, and Monte Carlo searches. The software documentation should be checked to ensure that the method used will preserve chirality. Some of the simplest Monte Carlo algorithms tend to invert chiral centers. As insertions tend to be 1–10 residues, CPU time requirements are usually not a concern.

### 9.2.6 Step 6: Optimization of the Model

The geometry of the protein obtained by following the steps up to this point in the procedure is usually not the optimal geometry. The backbone geometry taken from the template should be close to the backbone geometry of the unknown, but is probably not identical. When the backbone geometry is adjusted, that in turn may make it necessary to readjust side chain geometries, even changing to different rotamers.

The tricky part is choosing an optimization scheme that is appropriate. Energy minimization often improves the backbone position, but it cannot shift side chains from one rotamer to another. Conformation search techniques tend to explore many conformations that are radically different from the starting point, but the value of homology modeling is that the correct solution should be very close to the position obtained from the template.

One solution to this quandary is to run a molecular dynamics simulation. By adjusting the temperature, it is possible to give the simulation enough energy to explore the nearby conformational space, without unraveling the folds of the protein. Another option is to use various types of constrained conformation search.

There are cases where the accuracy of the homology model appears to be limited by the accuracy of the energy evaluation method used in the optimization. Therefore, some investigation into the accuracies of various force fields is justified. Note that it may also be necessary to include explicit solvent molecules, as some proteins require structural water molecules in order to fold correctly.

### 9.2.7 Step 7: Model Validation

Once the homology model has been constructed, it is highly advisable that the researcher take all reasonable steps to check whether this has been done correctly. There is no absolutely reliable way to do this, other than comparison with the crystal structure. However, there are a number of metrics that can suggest whether the model is likely to be good or bad. The following paragraphs discuss these options.

Note that force field energy is not an indicator of a good homology model. Cases have been documented in which a protein was completely misfolded, but well optimized. In these cases, the total energy from the molecular mechanics minimization was nearly identical to the total energy for the correctly folded structure.

General checks for reasonable bond lengths, bond angles, and torsion angles are often used in crystallography as a "sanity check." In homology model building, these give just a crude check for very severe problems. Most of the time, if the structure has been minimized with molecular mechanics, the minimization will ensure that these parameters are reasonable. If a bond angle or torsion angle is still unreasonable after molecular mechanics minimization, this often means that a side chain is in the wrong rotamer. However, very badly misfolded structures can sometimes pass this check.

A visual representation of how reasonable the conformational rotations are can be given by a Ramachandran plot, as shown in Fig. 9.4. This shows whether the distribution of backbone bond angles is similar to that typically found in correctly folded proteins. Ramachandran plots and other statistical measurements can be generated with the PROCHECK program, and many other protein analysis software packages. The Verify 3D program computes the probability of a side chain occupying a specific region. The ERRAT program examines the distribution of nonbonded atom–atom interactions for key atoms. The ProSa program uses the potential of mean force. The PROVE

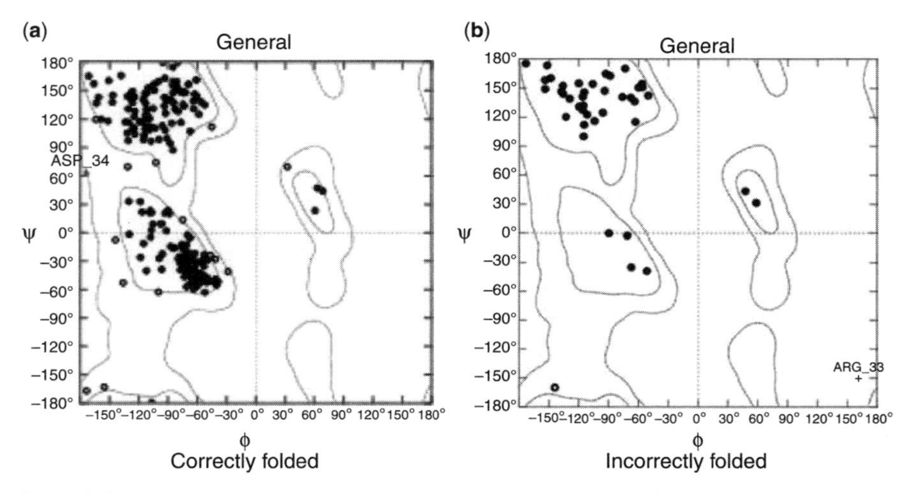


**Figure 9.4** Examples of Ramachandran plots: (**a**) for a correctly folded protein (caspase-3); (**b**) for an incorrectly folded protein from the CASP entries.

software looks at computed volumes. Most researchers use a number of these programs, and more. Structures or regions of structures that fail some or all of the checks are "red-flagged" for more careful examination.

The inside/outside distribution of hydrophilic and lipophilic residues is one of the better, frequently used checks. A protein that is immersed in solvent in its native environment will tend to have hydrophilic residues at the exterior of its globular structure, and lipophilic residues at the interior of the structure. There will often be some exterior lipophilic residues and some interior hydrophilic residues, but they should be the minority. If the protein is natively embedded in the cell membrane, it will tend to have hydrophilic residues on the outside at the solvent-exposed ends, and lipophilic residues around the middle where they are adjacent to the lipid bilayer. Some software packages have an option to color residues according to whether they are hydrophilic or lipophilic, which makes this very easy to analyze via a visual check.

Gregoret and Cohen published the QPACK algorithm, which was originally written in a freestanding software package, also named QPACK. This is an evaluation of whether the homology model is reasonable based on residue–residue packing density. Each residue is described by 1–3 spheres, and a bump check is done to see whether these spheres are too close together, an optimal distance, or further apart than expected. They also utilize a graphical display to color amino acids according to these three cases. The exterior residues tend to be indicated as being further away from neighboring residues, even though the geometry is just fine. An indication of interior residues being too close together means that it is likely that the model is not very good. Indications of interior residues being too far away might be a problem with the fold, or they may just be locations in which the protein should have structural water molecules hydrogen-bonded into the folded protein.

A useful, but somewhat more obtuse, check is to use radial distribution functions. These check the distance between every atom type and every other atom type. For example, there would be a distribution of distances between the sulfur in a methionine and the nitrogen in a tryptophan. This distribution can be compared with pretabulated distributions for known correctly folded structures. When two atoms are a distance where they seldom occur, it provides a "red flag" to the model builder that a particular region of the model should be looked at more closely. The software package ProsaII implements this technique.

A variation on the distribution function technique is to look at the direction of nearest-neighbor nonbonded atomic contacts. This is implemented in the WHAT_CHECK program, which is used for validation of crystallographic structures as well as computationally derived models.

The techniques discussed in this section give some tools for suggesting when a protein structure might be good or bad. The only definitive test is to compare the structure with a crystallographically determined structure.

As such, the researcher's experience and judgment are valuable tools that should not be overlooked in the model building process.

### 9.2.8 Step 8: If Errors Are Found, Iterate Back to Previous Steps

It is not unusual that a model fails validation, and steps of the process must be repeated. The step that should be returned to depends upon the nature of the validation failure.

If the inside/outside location of hydrophilic and lipophilic residues is incorrect, this means that the backbone itself is incorrect. In this case, the process may need to be started over, perhaps with a different template structure. Depending upon how well the early steps went, the researcher may choose to switch templates, or rework the alignment to the template. If the initial structure was based on a single template with marginal identity, one should consider using regions of high identity from several different template structures.

Sometimes, the backbone seems fine, but there are odd angles on the side chains. In this case, one should try different side chain rotamers for the residues in question. Reoptimizing the entire model with a different algorithm can also give improved results.

If the majority of the errors seem to be in the loops, one should try reoptimizing just those regions. Loops are often the source of the largest errors, even in experimentally determined structures.

Up to this point, this chapter has talked about homology model building in general. In practice, various software packages will come with a given set of tools intended to be used following a particular design process. The following are some notes on the homology building software packages on the market:

- The GeneFold and Composer modules from Tripos Inc. are tools for piecewise homology model building. They can be used to follow a fairly typical process of building a homology model from fragments, and optimization of loops and side chains. MatchMaker from Tripos uses fingerprint matching (called inverse folding) to give homology models when there is only a weak similarity.
- The MOE software from Chemical Computing Group also has a piecewise homology model generation. Theirs is more statistical. It makes up a whole ensemble of possible conformers in a database. Backbone fragments come from a database of crystal structures. Side chain orientations can be selected from a rotamer library. Finally, optimization is performed on each. All of these can be examined with various structural quality assessment tools.
- Accelrys has homology model generation tools available through their Discovery Studio interface. These are some of the more automated tools for generating a homology model from a single template structure.

The tools do this as a multistep process, to leave the option of making manual adjustments if desired.

- Schrödinger Inc. sells a protein structure prediction module called Prime. The Prime module uses threading and fold recognition. These allow it to find a best backbone template when none with the desire homology exist. The brochure says that these algorithms are to "create backbone models for early structural investigations or functional annotation," thus implying that it is not a high accuracy situation. The Prime module also includes tools for more conventional homology model building, which can give more accurate protein structures.

## 9.3 RELIABILITY OF RESULTS

Every few years, there is a CASP (Critical Assessment of Techniques for Protein Structure Prediction) competition. In this competition, researchers can generate models of proteins; the then judges determine how well each did. The crystallographic structure for the challenge sequence is known, but is withheld from publication until the groups have submitted their models. These competitions have three categories: Comparative Modeling (homology models), Sequence to Fold Assignment, and *Ab Initio* Folding. The results include both models created with the manual intervention of an expert scientist and models created by completely automated software packages.

The predictions submitted to CASP are evaluated with automated algorithms identified by acronyms such as GDT_TS, AL0_P, and DAL4. Some cutoff value of these metrics is chosen to indicate whether the prediction was successful. Typically, a structure is labeled a success if it contains the majority of the major fold features (i.e., $\alpha$-helices or $\beta$-sheets), although the loops may still be significantly in error. In the seventh competition (CASP7), the percentage of predictions that were deemed qualitatively correct ranged from 5% to nearly 100%, depending upon which protein was being examined. These automated tests are used because structures that fail tend to be so far away from the correct shape that it is impossible to perform an alignment, and thus impossible to compute a RMSD value. Thus far, none of the structure prediction methods has been shown to be consistently accurate. However, there are some general conclusions that can be drawn form the CASP results:

- Each method shows a very broad range of results, ranging from extremely accurate to completely wrong and qualitatively incorrect.
- On average, homology modeling gives more accurate results than *ab initio* protein folding.

- The best results from the completely automated software packages rival the accuracy of the best results from the models created with manual intervention.

Homology model building has two important advantages over protein folding. First, it is more accurate on average. Second, and more importantly, the researcher can get a better estimate of whether the homology model is likely to be qualitatively and quantitatively accurate, based on the degree of similarity to a known structure. The role and reliability of homology model building is increasing as the number of available crystal structures increases.

## BIBLIOGRAPHY

### Accuracy of Homology Models

Baumann G, Frommel C, Sander C. Polarity as a criteria in protein design. Protein Eng 1989; 2: 329–334.

CASP1 Proceedings. Proteins 1995; 23: 295–460.

CASP2 Proceedings. Proteins 1997; 29(S1): 1–230.

CASP3 Proceedings. Proteins 1999; 37(S3): 1–237.

CASP4 Proceedings. Proteins 2001; 45(S5): 1–199.

CASP5 Proceedings. Proteins 2003; 53(S6): 333–595.

CASP6 Proceedings. Proteins 2005; 61(S7): 1–236.

CASP7 Proceedings. Proteins 2007; 69(S8): 1–207.

Gregoret LM, Cohen FE. Novel method for the rapid evaluation of packing in protein structures. J Mol Biol 1990; 211: 959–974.

Holm L, Sander C. Evaluation of protein models by atomic solvation preference. J Mol Biol 1992; 225: 93–105.

Hooft RWW, Vriend G, Sander C, Abola EE. Errors in protein structures. Nature 1996; 381: 272–272.

Morris AL, MacArthur MW, Hutchinson EG, Thorton JM. Stereochemical quality of protein structure coordinates. Proteins 1992; 12: 345–364.

Novotný J, Rashin AA, Bruccoleri RE. Criteria that discriminate between native proteins and incorrectly folded proteins. Proteins 1988; 4: 19–30.

Rost B. Twilight zone of protein sequence alignments. Protein Eng 1999; 12: 85–94.

Sippl MJ. Calculation of conformational ensembles from potentials of mean force. J Mol Biol 1990; 17: 355–362.

Vriend G, Sander C. Quality control of protein models: Directional atomic contact analysis. J Appl Crystallogr 1993; 26: 47–60.

## Homology Modeling Reviews

Blundel TL, Sibanda BL, Sternberg MJE, Thornton JM. Knowledge-based prediction of protein structures and the design of novel molecules. Nature 1987; 326: 347–352.

Esposito EX, Tobi D, Madura JD. Comparative protein modeling. Rev Comput Chem 2006; 22: 57–167.

Fetrow JS, Bryang SH. New programs for protein tertiary structure prediction. Bio/Technology 1993; 11: 479–484.

Greer J. Comparative modeling of homologous proteins. Meth Enzymol 1991; 202: 239–252.

Johnson MS, Srinivasan N, Sowdhamini R, Blundell TL. Knowledge-based protein modeling. Crit Rev Biochem Mol Biol 1994; 29: 1–68.

Krieger E, Nabuurs SB, Vriend G. Homology modeling. In: Bouirne PE, Weissig H, editors. Structural Bioinformatics. New York: Wiley-Liss; 2003. p. 507–521.

## Homology Modeling Tutorials

Martz E. Homology Modeling for Beginners with Free Software. Available at http://www.umass.edu/molvis/workshop/homolmod.html. Accessed 2006 June 22.

Sali A, Overington JP, Johnson MS, and Blundell TL. From comparisons of protein sequences and structures to protein modelling and design. Trends Biochem Sci 1990; 15: 235–240.

Vriend G. Professional Gambling. Available at http://swift.cmbi.ru.nl/gv/articles/text/gambling.html. Accessed 2006 July 13.

Additional references are contained on the accompanying CD.

# 10

# MOLECULAR MECHANICS

There are many good books on molecular mechanics. Therefore, this chapter is not going to try to be a thorough primer on the subject. Instead, we will discuss some of the aspects of molecular mechanics, as it relates to drug design work. Section 10.1 will give only the briefest description of molecular mechanics, in order to set the stage for the discussion of how it is utilized in drug design.

## 10.1 A REALLY BRIEF INTRODUCTION TO MOLECULAR MECHANICS

On the off-chance that the reader of this book is not well versed in molecular mechanics, this section of the chapter is provided to establish some basic terminology. Molecular mechanics, like quantum mechanics or semiempirical methods, is a generic method for the simulation of molecules. At its heart, molecular mechanics provides an equation for computing the energy of a molecule. With the use of this energy expression, various algorithms can be used to determine the preferred shape of a molecule, the energetics of its interaction with other molecules, and the way in which it can move.

The energy computed by a molecular mechanics calculation is a conformational energy. Thus, it is not an energy of formation, successive ionizations,

*Computational Drug Design*. By David C. Young

or any other experimentally measurable energy. However, energy differences are computed reasonably accurately using molecular mechanics. This means that the method will usually accurately compute the difference in energy between different conformers, and other relative energetics due to the geometric position of the atoms.

The scientifically correct description of the energetics of a molecular system is given by the Schrödinger equation of quantum mechanics. However, quantum mechanical calculations are too time-consuming to use on biochemical species, because of the necessity of modeling every electron in every orbital, most of which are described by multiple basis functions. Rather, molecular mechanics is a computationally simpler technique that uses an algebraic formulation in place of the partial differential equation formulation used in quantum mechanics.

The molecular mechanics energy equation is a sum of terms that calculate the energy due to bond stretching, angle bending, torsional angles, hydrogen bonds, van der Waals forces, and Coulombic attraction and repulsion. The bond stretching term is most often harmonic, but there are also anharmonic potentials such as the Morse potential. Angle bending terms are almost always harmonic, and torsional angle terms are usually cosine functions. All of the parameters in these equations are fitted to experimental data. A set of equations and the parameters that go with them is called a "force field." Most force fields are parameterized to describe organic molecules, and a few can model inorganic atoms.

Nearly all of the force fields are designed to model molecular structures near equilibrium only. As such, most can model molecules and solvent–solute systems. Most force fields cannot model chemical reactions. Specifically, the formation and breaking of chemical bonds are not described by the equations (i.e., by harmonic potentials), either quantitatively or qualitatively. Most force fields include partial charges on the atoms, but are not designed to model charge transfer reactions.

Molecular mechanics methods are the basis for other methods, such as construction of homology models, molecular dynamics, crystallographic structure refinement, and docking. Understanding the strengths and weaknesses of various force fields can help to understand how to most effectively utilize all of these techniques.

There are two ways of putting solvation into a force field calculation. One option is to build a collection of water molecules to give an explicit solvation. The second option is to put in a continuum solvent description defined by a dielectric constant. For practical reasons, the continuum method is usually used, unless the situation requires explicit solvent molecules.

The advantages of using explicit water molecules are that it can give a more accurate description, and can find positions where a structural water molecule must be in the protein. The disadvantage is that each energy calculation takes more CPU time. This is multiplied many times over, in that a molecular dynamics or Monte Carlo calculation must be performed to obtain the statistical average of many solvent molecule positions.

The advantage of the continuum solvation energy expression is that the calculations run much faster. The solvent dielectric should be included only outside the protein, or in gaps large enough for solvent molecules to be present. The dielectric constant of bulk water is approximately 78, but may have an effective value of 2–5 in the protein core.

## 10.2 FORCE FIELDS FOR DRUG DESIGN

No force field is perfect. Every force field will give more accurate results for some chemical systems, and less accurate results for others. The functional form of the energy terms selected by the force field creators is typically the minor contributor to these differences in accuracy. The major contributor to the accuracy is typically the choice of data used to parameterize the force field. Table 10.1 lists some common force fields and notes the original designed usage.

Another consideration in force field selection is robustness. A less robust force field may give very accurate results for one chemical system, but very poor results for another. A robust force field will give good results for a wide range of chemical systems, even if it is not always the most accurate. The CFF, MMFF, MM2, MM3, and MM4 force fields are sometimes classified as second-generation force fields because they incorporate improvements to the energy expression and parameterization algorithm that serve to improve robustness while maintaining good accuracy.

Molecular mechanics methods are used for many different types of calculations. The method by itself computes the energy associated with a given conformation, which can in turn be used to determine the optimal shape of the molecule. Molecular dynamics calculations incorporate Newton's equations of motion to simulate molecular vibration and movement within a solvent. Monte Carlo methods use a random number generator to select many molecular positions in order to obtain a statistical description of the system. Docking calculations determine the best orientation and conformation for an inhibitor to bind in a protein's active site. There are many more examples.

Commercial drug design software is often sold as a collection of modules (small programs), each of which is designed to perform a single, specialized task. Typically, all of the modules sold by a given manufacturer will work within a common graphical interface. Many of these specialized modules are designed to use one specific force field, even though other force fields

**TABLE 10.1 Force Field Primary Usage**

| Force Field | Primary Use |
|---|---|
| AMBER | Proteins and nucleic acids |
| AMOEBA | Proteins |
| CHARMM | Proteins and nucleic acids |
| CFF | Protein vibrational motion |
| UBCFF, CVFF, QMFF, CFF93 | Variations of CFF |
| CHEAT | Carbohydrates |
| CPE | Molecular liquids |
| DREIDING | Organic and bio-organic molecules |
| DRF90 | Solvent–solute molecular dynamics |
| ECEPP | Peptides |
| EFF | Hydrocarbons |
| ENZYMIX | Modeling chemical reactions in biological molecules |
| EVB | Modeling chemical reactions in different environments |
| GROMACS | Molecular dynamics |
| GROMOS | Molecular dynamics of solvent–solute systems |
| MM1, MM2, MM3, MM4 | Organic molecules |
| MMX, MM+ | Variations of MM2 |
| MMFF | Organic and bio-molecular systems |
| MOMEC | Transition metal coordination compounds |
| OPLS | Liquid simulations |
| OPLS-AA, OPLS-UA, OPLS-2001, OPLS-2005 | Derivatives of OPLS |
| PFF | Molecular dynamics |
| QCFF/PI | Conjugated molecules |
| ReaxFF | Dynamical simulations of chemical reactions |
| RFF | Simulation of chemical reactions, full periodic table |
| SIBFA | Small molecules and flexible proteins |
| SYBYL | Another name for the Tripos force field |
| Tripos | Organic and bio-organic molecules |
| UFF | Full periodic table |
| YETI | Nonbonded interactions, such as docking |

may be available in the program. In this case, it is best to follow the manufacturer's recommendation, unless an alternative has been validated very carefully.

For general tasks, such as optimizing chemical structure, there are usually a number of force fields that will give good results. Often, researchers working in a particular research group or on a particular type of chemical system will prefer one force field, based on experience with results for that system. Of the force fields discussed in this chapter, the MMFF, CHARMM, AMBER, Tripos, and OPLS force fields seems to be utilized more often than others for drug design applications. Each of these has been utilized for a very wide range of chemical systems, particularly complexes between drugs and proteins. The MMFF force field may be the most widely used of these, as it has been integrated into a number of different software packages, and gives consistently reasonable results across a wide range of chemical systems. The others are more often used in conjunction with software created by the companies that market those methods.

A reasonable understanding of molecular mechanics is important, because this can give insights into the strengths and weaknesses of many techniques built on molecular mechanics. If used properly, molecular mechanics can be an accurate and reliable workhorse.

## BIBLIOGRAPHY

### Books

Allen MP, Tildesley DJ. Computer Simulation of Liquids. Oxford: Clarendon Press; 1997.

Leach AR. Molecular Modelling Principles and Applications. Harlow, Essex, UK: Longman; 1996.

Rappé AK, Casewit CJ. Molecular Mechanics Across Chemistry. Sausalito, CA: University Science Books; 1997.

Young D. Computational Chemistry: A Practical Guide for Applying Techniques to Real World Problems. New York: Wiley; 2001.

### Review Articles

Gilson MK, Honig BH. The dielectric constant of a folded protein. Biopolymers 1986; 25: 2097–2119.

Gundertofte K, Liljefors T, Norrby PO, Pettersson I. A comparison of conformational energies calculated by several molecular mechanics methods. J Comput Chem 1996; 17: 429–449.

Pérez S, Imberty A, Engelsen SB, Gruza J, Mazeau K, Jiménez-Barbero J, Poveda A, Espinoza JF, van Eijck BP, Johnson G, French AD, Kouwijzer MLCE, Grootenhuis PDJ, Bernardi A. Raimondi L, Senderoxitz H, Durier V, Vergoten G, Rasmussen K. A comparison and chemometric analysis of several molecular mechanics force fields and parameter sets applied to carbohydrates. Carbohydr Res 1998; 314: 141–155.

Whitnell RM, Wilson KR. Computational molecular dynamics of chemical reactions in solution. Rev Comput Chem 1993; 4: 67–148.

Zimmer M. Bioinorganic molecular mechanics. Chem Rev 1995; 95: 2629–2649.

Additional references are contained on the accompanying CD.

# 11

# PROTEIN FOLDING

There are circumstances where a researcher wants to use structure-based drug design techniques, but there is no crystal structure of the target, any 2D NMR data, or a crystal structure for a sufficiently similar protein to make homology model building reliable. In this case, the remaining option is to try computationally to find a complete geometry from the primary sequence data. This is called a protein folding calculation.

Some years back, there was a flurry of protein folding research. These research projects resulted in a selection of available algorithms and software packages. However, as the previous paragraph implies, protein folding is often an option of last resort. Protein folding can give very accurate protein structures, but it can also give very inaccurate structures. The problem is that it is very difficult to determine whether the structure obtained accurately reflects the true three-dimensional shape of the protein. Many studies have also been carried out to give an understanding of the pathway and kinetics by which a protein actually becomes folded in the body, but this application will not be covered in this chapter.

## 11.1 THE DIFFICULTY OF THE PROBLEM

In principle, protein folding is just very computationally intensive. In practice, there are far more problems to be overcome.

The basic idea of a protein folding calculation is that many possible conformers of the protein can be computed, and the lowest energy one will be the folded ground state of the molecule. However, the assumption that the lowest energy conformer is the correct folded geometry is not necessarily true. There have been documented cases in which a completely misfolded but well-optimized protein structure had a total energy nearly identical to that of the correctly folded structure. Thus, computational techniques may not always be sufficiently accurate to correctly predict which structure should have the lowest energy. The energy is certainly dependent upon the presence of a solvation shell around the protein, structural water molecules folded inside the protein, vibrational energy levels, and disulfide bond formation. If the protein being folded is membrane-bound, it may be necessary to include a lipid bilayer in the calculation.

There are also proteins that are folded by chaperones. In this case, the lowest energy conformer assumption is certainly wrong, even if the computational method is sufficiently accurate to find this conformer. The chaperones serve to force the protein into a conformation that is a local minimum, but is not the lowest energy global minimum. Thus, the only way to determine the structure of a chaperone-guided folding completely theoretically would be to compute the entire folding pathway in the presence of the chaperone.

One analysis suggested that the accuracy necessary to find the global minima is on the order of 1 kcal/mol per residue. It may not be feasible to use a technique where the maximum error is within this margin, but one would at least hope for a technique where 1 kcal/mol was perhaps three standard deviations above the average error. Most of the energy computation techniques known do not reach this lofty goal—certainly not if they are capable of computing energies for large biomolecules with a reasonable amount of memory and CPU time.

Once issues of accuracy have been addressed, the issue of very computationally intensive algorithms must be addressed. Consider the simplest grid search, in which each rotatable bond in the protein is tested in each of the three staggered conformations. Computing all combinations of each bond in each of these three conformations, a protein with 1800 rotatable bonds would require $3^{1800}$ structures to be computed. If this computation were to be done at the optimistic rate of one million structures per second, the entire computation would take approximately 112 years to complete!

At first glance, these concerns make the determination of protein folding seem like an impossible task. However, a number of algorithms have been designed specifically for this purpose. These specialized programs make it possible, albeit rather difficult, to obtain a folded protein structure starting from the primary sequence. The next section describes some of these algorithms.

## 11.2 ALGORITHMS

A very large number of protein folding algorithms have been proposed. It is not practical to explore all of the subtle variations on these algorithms in this text. However, the following is a discussion of each general type of algorithm. The discussion of advantages and disadvantages of each is based on the original, simplest form of that algorithm. Note that some of these issues may not be applicable for specific implementations, which may use modified forms of the algorithms. It is advisable to consult the literature reference for the specific implementation being used before applying it to a research problem.

The timing example above is based on a grid search algorithm. Grid searches utilize the very simplistic view of checking every conformer with every bond at some chosen set of torsion angles. This is useful for determining the lowest energy conformers of rather small molecules. However, this algorithm is useless for any sizeable protein because of the $X^Y$ time complexity ($Y$ bonds checked at $X$ different torsion angles).

A Monte Carlo search samples the space of possible conformers, but does not compute every single conformer. It randomly chooses conformation angles to generate each structure. Often, a list of the lowest energy structures is maintained, and then each of these is optimized with a simple downhill energy optimization. Monte Carlo searches generally give a near-optimal solution much faster than a grid search. Many variations on the Monte Carlo algorithm have been proposed, such as Tsallis statistical Monte Carlo, jump-walking Monte Carlo, multicanonical Monte Carlo, and configuration bias Monte Carlo. Many of these variations differ in the way in which the move is chosen. A good overview of these variations is in Straub's review article, referenced at the end of this chapter.

Simulated annealing can be viewed as an improved variation of the Monte Carlo search. Note that there are several different formulations of simulated annealing. In all of them, the energy-accessible solution space is reduced once the calculation is far enough along to have started finding some low energy regions. This is sometimes done by reducing the temperature in a statistical mechanical simulation, which might be a Monte Carlo simulation or sometimes a molecular dynamics simulation. Another form of simulated annealing starts with a Monte Carlo search, and then gradually changes the statistical probability to make conformers near the lowest energy conformer thus found far more likely to be generated in subsequent steps.

Many molecular dynamics-based searches have also been utilized. Molecular dynamics simulations tend to sample fewer of the higher energy conformations than Monte Carlo, at the expense of having a tendency to get caught in local minima. In order to address these issues, several noisy or stochastic molecular dynamics algorithms have been developed, which add randomness to the molecular dynamics simulation. Some techniques also modify the potential energy surface in order to make the space easier to search.

A tabu search is another variation of a Monte Carlo search. In a tabu search, information about the regions of conformation space already sampled is stored in memory. This information is used to ensure that the search does not spend time computing conformers that are very similar to those already found to have poor energies. Thus, a tabu search is often equivalent to a Monte Carlo search with significantly more iterations.

Another basic type of algorithm is a chain growth algorithm. In this, not all of the torsion angles are changed at once. The structure is built up one residue at a time, with the conformers of the bonds only being checked in the residue last added. This explores a reasonable set of potential conformers in a short amount of time. Multiple chain growth runs must be carried out in order to give a reasonable sampling of conformer space.

The ant colony search runs multiple chain growth runs in a single process. This search samples the surrounding low energy space efficiently. It does this by using some of the conformation angles from previous runs (analogous to ants following a pheromone trial part way along its path).

A piecewise optimization (or incremental folding) is somewhat similar to a chain growth search in that it does not change all of the conformation angles at once. In a piecewise search, part of the structure is optimized while part of it is held fixed. Often, the backbone is optimized independently from the side chains. The pieces are then sometimes moved around relative to one another after creating a potential to quantify the interactions between the pieces.

Knowledge-based searches (or heuristic searches) have a similar strategy of optimizing one section at a time. In a knowledge-based search, the primary sequence is compared with known sequences in an attempt to identify sections that will fold into known conformations, such as α-helices or β-sheets. These regions are then held fixed while the rest is optimized. The philosophy of this type of search straddles the definitions of protein folding and homology model building.

Genetic algorithms are based on the way in which genetics tends to optimize animal traits for survival in their environment. Many trial copies of the problem (protein conformers) are optimized at once. The optimization functions include characteristics found in nature, such as increased survival rates for the fitter individuals, successive generations that combine traits of the parents, and random mutations. As a general rule, genetic algorithms work well if all of the possible factors are taken into account, but poorly if a simpler form of the algorithm is used. The poor performance of simplified genetic algorithms is usually due to premature convergence to a set of nonoptimal traits.

Some papers present protein folding algorithms that place atoms on a square or triangular grid. This can be a useful simplification for algorithm development. However, grids should not be used when actually attempting to generate an accurate three-dimensional structure.

Because protein folding calculations are so computationally intensive, the software may be written to use multiple CPUs on a computing cluster. This idea has been taken to the extreme by projects such as folding@home, which utilize idle time on tens of thousands of computers around the world.

When choosing a software package for protein folding, it is advisable to look up the developer's published papers. Did they use a grid? Did they test it on small proteins or proteins of the sizes found in the body? Was it tested on a membrane-bound protein?

## 11.3 RELIABILITY OF RESULTS

Chapter 9 listed a number of tests that can be used to give an indication of whether the homology model building process produced a reasonable structure. With the exception of utilizing similarity to the template sequence, these tests can also be applied to protein folding results. The test used most

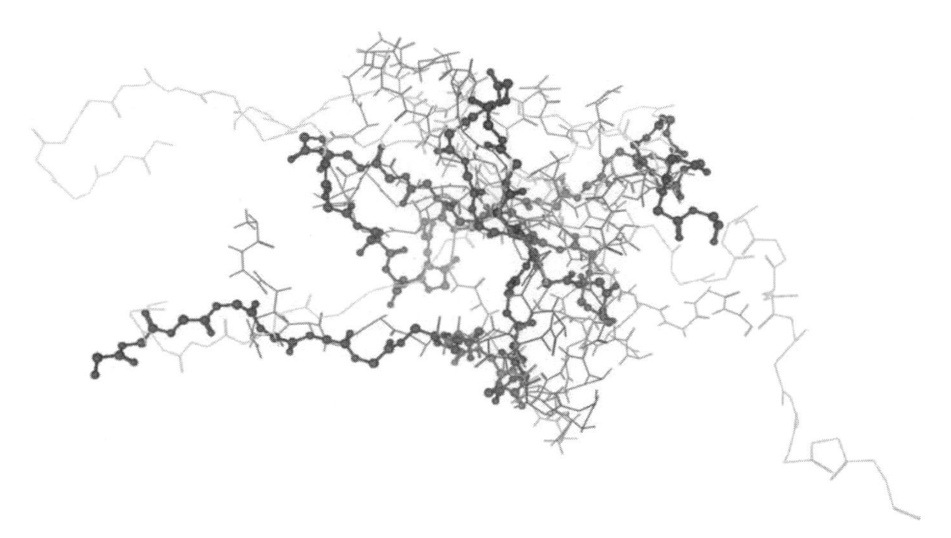

**Figure 11.1** A few structures selected at random for target T0348 from the CASP7 competition. One found the β-sheet, but others were not even close to the known structure (rendered as a ball and stick model). A little over half of the predictions submitted for this target were found to be qualitatively correct.

frequently is to check whether the hydrophilic residues are on the outside of the structure and the lipophilic residues on the inside.

The CASP results mentioned in Chapter 9 include protein folding results as well. They show that it is possible for protein folding calculations to give an accurate structure. However, homology models give results that are more often reasonable and more accurate. It is no accident that the most successful protein folding algorithms are those that are most similar to homology model algorithms—that is to say, those in which prior analysis of protein structures is used to assign backbone conformations. Figure 11.1 shows an example of some results from the CASP competition.

Protein folding has received much attention in the literature. However, it has not been a success in the pharmaceutical industry. It is an elegant idea, but most consider it too risky a basis upon which to spend millions of dollars on a drug design project. Indeed, industrial research projects may be cancelled if protein folding is the only option.

## 11.4 CONFORMATIONAL ANALYSIS

The same conformation-searching algorithms that can be used for protein folding can also be used for finding the lowest energy conformer or all conformers of a small molecule. When used for small molecules, this is typically called conformational analysis. The major difference between the two applications is that the conformers of a small molecule can be exhaustively analyzed with a reasonable amount of computer power, whereas protein folding can be an intractable problem.

Conformational analysis is often done en masse to facility the analysis of a large database of compounds. For example, some pharmacophore searching tools look only at the conformer stored in the database, thus making it necessary to preprocess the database to contain the conformers of interest. In some cases, only the lowest energy conformer of a compound is desired for subsequent analysis. In some cases, several conformers are kept. It is possible to analyze all conformers of each compound, but this can be prohibitive owing to the excessively large number of compounds and amount of time necessary to analyze them. Perhaps the most thorough option that is reasonable is to keep all of the biologically accessible conformers, which are the lowest energy conformer and any conformer within 10 kcal/mol of the lowest energy.

Conformational analysis programs may also have a function for aligning chemical structures. This can be necessary for tasks such as pharmacophore detection and 3D-QSAR model building.

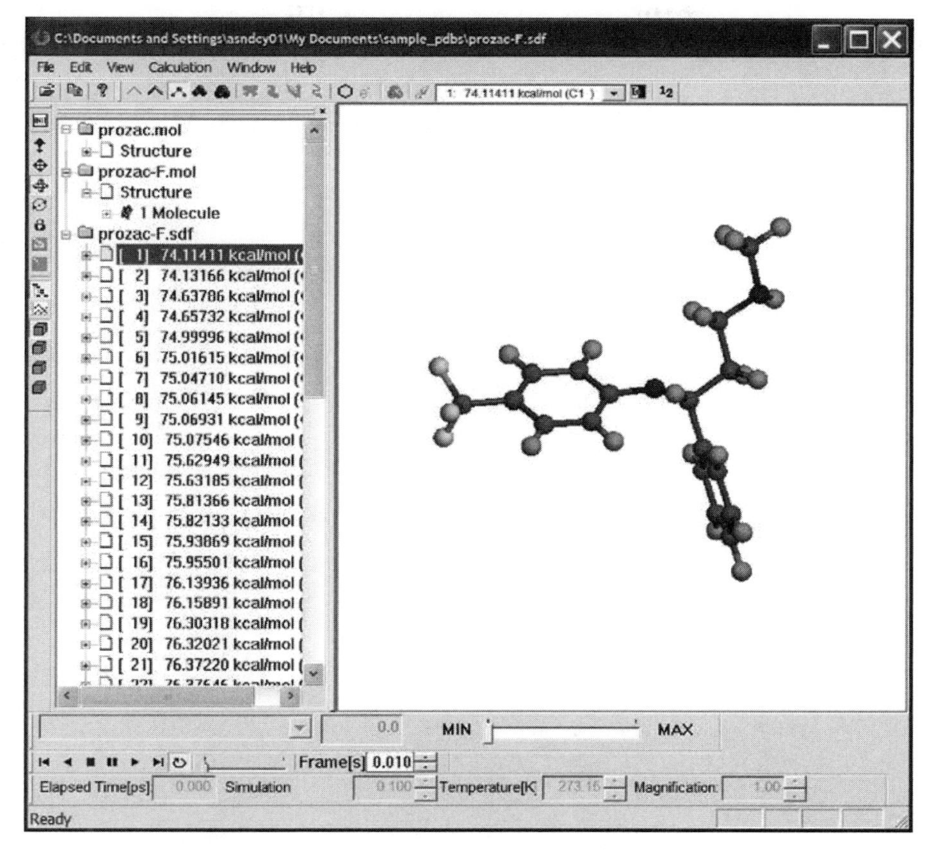


**Figure 11.2** CONFLEX is a stand-alone program for conformational analysis.

In many cases, conformational analysis is built into full-featured drug design software suites. There are also stand-alone conformational analysis programs, such as CONFLEX and its BARISTA graphical interface shown in Fig. 11.2.

Protein folding research has yielded some valuable algorithms for analysis of small molecules. However, it has not proven to be a sufficiently reliable mainstream technique for determining structures of large proteins.

## BIBLIOGRAPHY

Friesner RA, editor. Computational Methods for Protein Folding (Advances in Chemical Physics. Volume 120). New York: Wiley; 2002.

Godzik A, Kolinski A, Skolnick J. *De novo* and inverse folding predictions of protein structure and dynamics. J Comput Aided Mol Des 1993; 7: 397–438.

Straub JE. Protein folding and optimization algorithms. In: von Ragué Schleyer P et al., editors. Encyclopedia of Computational Chemistry. New York: Wiley; 1998. p. 2184–2191.

# 12

# DOCKING

## 12.1 INTRODUCTION

Docking is an automated computer algorithm that determines how a compound will bind in the active site of a protein. This includes determining the orientation of the compound, its conformational geometry, and the scoring. The scoring may be a binding energy, free energy, or a qualitative numerical measure. In some way, every docking algorithm automatically tries to put the compound in many different orientations and conformations in the active site, and then computes a score for each. Some programs store the data for all of the tested orientations, but most only keep a number of those with the best scores. Docking functionality is built into full-featured drug design programs, and sold as stand-alone programs, sometimes with their own graphical interface. Figure 12.1 shows a typical stand-alone docking program graphical interface.

Docking is probably the most heavily used tool in computational drug design. It is also the most accurate method for predicting whether a particular compound will be a good inhibitor of a particular protein. For this reason, pharmaceutical companies analyze very large numbers of compounds with docking. Those compounds that have the best docking results will be synthesized if necessary, and analyzed in the laboratory. The large lists of compounds that are analyzed computationally may be of compounds designed in the computer, compounds that are available for purchase, or compounds

*Computational Drug Design.* By David C. Young

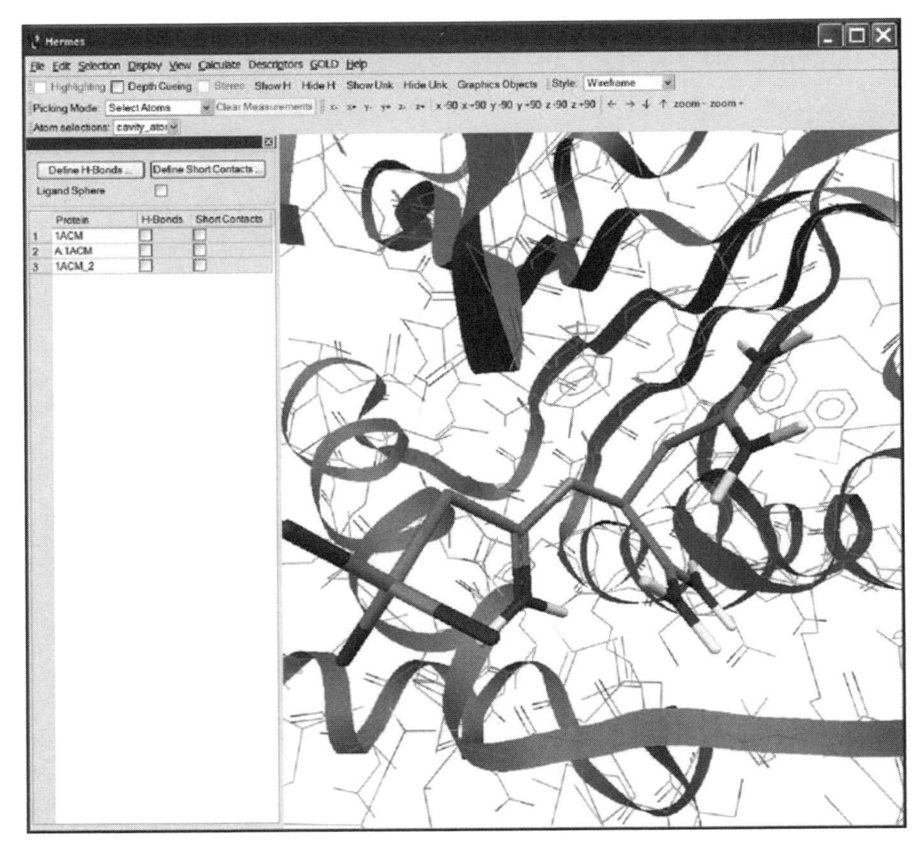


**Figure 12.1** GOLD is a docking program available from the Cambridge Crystallographic Data Centre. It is one of the docking programs that are widely used in the pharmaceutical industry.

in the company compound library. By using the knowledge gained from the docking study, fewer compounds need be synthesized and assayed, and a higher percentage of compounds assayed are found to be active.

Because docking calculations simulate the interaction between a compound and a protein's active site, the results are comparable to those of biochemical assays. Most scoring functions compute some sort of energy. Most biochemical assays compute an inhibition rate constant $K_I$. Thus, the binding energy from the docking calculation should be proportional to ln $K_I$ from biochemical assays (a simple Arrhenius equation relationship). Some scoring functions show good correlation to ln $K_I$, while others give only a qualitative ranking showing which compounds are better or worse.

Docking results do not show a high degree of correlation to cell culture assays, efficacy in animals, or other tests. Like biochemical assays, the trends shown by docking results may be qualitatively similar to the trends in these other experiments. However, docking does not take into account bioavailability, toxicity, and other factors present in the body.

The primary reasons for using docking are to predict which compounds will bind well to a protein, and to see the three-dimensional geometry of the compound bound in the protein's active site. One limitation of docking is that a 3D structure of the target protein must be available. Also, the amount of computer time required to run docking calculations is not insignificant. Thus, it may not be practical to use docking to analyze very large collections of compounds. Less CPU-intensive techniques, such as pharmacophore or similarity searches, can be used to search very large databases for potentially active compounds. Compounds identified by those techniques are often subsequently run through a docking analysis. Pharmacophore searches are used to search databases of millions of compounds. Docking might be used to analyze tens or hundreds of thousands of compounds over the course of a multiyear drug design project.

## 12.2 SEARCH ALGORITHMS

### 12.2.1 Searching the Entire Space

There are two key components of a docking program: the search algorithm and the scoring algorithm. The search algorithm positions molecules in various locations, orientations, and conformation within the active site. Some of the earliest docking programs positioned a molecule in the active site, holding it rigid with respect to conformational changes, but all modern docking algorithms include ligand conformational changes. The choice of search algorithm determines how thoroughly the program checks possible molecule positions, and how long it takes to run.

Note that the search algorithm does not determine whether the docking program gives accurate results. The scoring function is responsible for determining if the orientations chosen by the search algorithm are the most energetically favorable, and is responsible for computing the binding energy. Thus, a search algorithm that does not sample the space thoroughly will give inaccurate results if the correct orientation is not sampled. However, most search functions will sample the space adequately if they are given the correct input parameters, thus making the accuracy of the results primarily dependent upon the accuracy of the scoring function.

In the following discussion, we will use several terms: "orientation," "translation," "conformation," and "position":

- *Conformation* has the typical definition of the 3D shape of the ligand being docked.
- *Translation* refers to a position in Cartesian space, relative to the protein's active site, which can be considered as a fixed shape for the discussion in this section.
- *Orientation* refers to rotations around the Cartesian axes while keeping the center of mass at a fixed point in space
- *Position* is used here to encompass every detail of the ligand's shape and location, including orientation, translation, and conformation.

The following is a discussion of some of the search algorithms presently in use, and their important features.

A Monte Carlo search is built around a random number generator. In the simplest implementation, position, orientation, and conformation are all chosen at random. Sometimes, position and conformation are checked independently. Thus, a position is chosen (translation), and many conformations are tested while in that position; then a new position is chosen, and the process repeats. Some Monte Carlo searches only change conformations of the rotatable bonds (single bonds not in rings), whereas others will reposition ring atoms. Repositioning ring atoms is a "double-edged sword," in that it can find different ring conformations, but may also invert stereochemistry. Monte Carlo algorithms can sample a space very thoroughly. However, modifications of the Monte Carlo algorithm, such as a tabu algorithm or a simulated annealing algorithm, often give the same accuracy of results with fewer iterations, and thus a faster run time.

Most tabu searches are implemented as a modified version of a Monte Carlo search. Like the Monte Carlo search, the tabu search chooses orientations and confirmations at random. However, the Monte Carlo algorithm utilizes no knowledge of what positions have already been sampled, and thus sometimes results in recomputing positions that have already been computed. The tabu algorithm keeps track of which positions have already been sampled, and avoids sampling those positions again. Thus, the tabu algorithm can give the same results with fewer iterations, by eliminating any duplication of work.

Simulated annealing algorithms are similar to tabu algorithms. Simulated annealing uses information about what positions have already been sampled. As the calculation progresses, positions that are near those that have already shown to have low energies are sampled more thoroughly. Positions that are near positions that previously gave high energies are sampled less thoroughly.

Again, this gives an accuracy of results similar to a Monte Carlo algorithm, but in fewer iterations. Simulated annealing algorithms and tabu algorithms often require somewhat more computer memory than Monte Carlo algorithms.

Genetic algorithms follow a different strategy. They are modeled after the process by which optimal genetic traits are obtained in a population through inheritance, mutation, elitism, etc. Thus, the calculation may start by generating a population (each member of which is a particular ligand position) at random. Then successive generations of molecules are generated by keeping mostly the best ("survival of the fittest") and combining traits from parents and random mutations. After several generations, the most fit (best scoring energy) should be well optimized. Genetic algorithms can sample a space thoroughly, if the parameters are chosen wisely, and can run very quickly. Genetic algorithms can require more memory than the other docking algorithms, as the entire population is often held in memory.

Most of the time, the implementation of these algorithms includes a bump check. This is a check that determines if two atoms are too close together. If this check fails, the orientation is discarded without going on to compute a score. Many of these algorithms also include at least a few steps of downhill minimization from each sampled position in order to obtain more accurate optimized positions. Some hold rings rigid, while others allow for ring inversions. Most prevent inversions of stereocenters.

Upon starting a large docking project, the researcher should run some validation tests to make sure that the parameters being used give acceptable results. This might mean setting the number of iterations, size of the area to be searched, grid size, or any other adjustable parameter. One reason that docking simulations can give incorrect results is insufficient torsional sampling. In many programs, this can be fixed by increasing the number of iterations in the search. The necessary number of iterations will be larger for larger molecules, more flexible molecules, or larger active sites.

### 12.2.2 Grid Potentials versus Full Force Field

Many scoring functions are simply molecular mechanics energies that only include interactions between the ligand and protein. Some also include torsional strain within the ligand as well. Selecting all pairs of atoms in a space requires $n^2$ evaluations for $n$ atoms. This can result in taking a considerable amount of CPU time to evaluate scoring energies, when the full force field equations are used.

One way to get the same answer much more rapidly is to create a potential field, which is usually evaluated numerically on a rectangular grid in the active site. The value of the potential at a given point on the grid is the energy of placing a unit charge at that point on the grid. The energy associated with putting a given atom at a particular point on the grid is then the potential at that point times the atom's charge (partial charge, not atomic charge). The grid, or a second one, may also represent steric hindrance or van der Waals forces. An extrapolation can be used when atoms are located between the grid points.

If the grid is properly constructed, a grid-based algorithm will give the exact same results as a full force field method, but will require an order of magnitude less CPU time to perform the docking calculation.

### 12.2.3 Flexible Active Sites

All docking programs alter the orientation and conformation of the ligand in order to fit in a protein's active site, which is itself held in a fixed geometry. Some docking programs have the ability to allow the active site to alter shape as well. This invariably results in docking simulations that take much longer to run. However, it is necessary to allow active site flexibility if the protein exhibits an induced fit mechanism. Even without an induced fit mechanism, it can give somewhat more accurate results owing to the residue side chains being able to reposition slightly.

There are several types of induced fit mechanism. Among the simplest is one in which the protein backbone stays rigid, but one or more of the protein residue side chains changes conformation to alter its position within the active site. A more complex mechanism is one in which a loop of the protein changes conformation to fold down on top of the active site, thus locking the substrate in place. The largest changes are seen in some proteins that exhibit a "clam shell" motion in which two halves of the folded protein separate, similar to a clam opening, in order to allow the substrate into an active site that is completely encapsulated when the halves close again. For example, cytochrome P450 enzymes are known to undergo massive conformational changes in order to accept a diverse range of ligands.

Many docking algorithms were originally developed to simulate the ligand binding in a crevice in the surface of the folded protein. Some programs have difficulty docking compounds in an active site that is completely enclosed. This can happen when the protein folds down over the active site or the entire active site opens and closes via a clam shell movement of two large sections of the protein. When this occurs, the researcher should look carefully at the software user manual. There are often optional inputs that will allow the program to function correctly with an encapsulated active site.

### 12.2.4 Ligands Covalently Bound to the Active Site

There are some inhibitors that form a covalent bond with one of the protein residues. This might be a reversible reaction, or an irreversible bond formation in the case of a suicide inhibitor. This situation creates a problem for docking studies. Most docking programs do not have any facility to allow a covalent bond to be formed between the ligand and active site.

A few docking programs have a way to add a constraint to the docking simulation. This keeps the ligand attached at the correct residue, and with the correct bond length. However, it does not result in the covalent bond energy being included. Since covalent bonds are much stronger than hydrogen bonds, this introduces a significant error in the scoring energy calculation. One option is to add a correction factor after the fact to include the covalent bond energy, usually obtained from a quantum mechanical calculation. Since different ligands often give different covalent bond strengths, this gives a significant improvement to the binding energy.

The ideal scenario for covalently bonded ligands would be to use a method that can accurately predict bond dissociation energies, that is, an *ab initio* method. This brings up the problem that *ab initio* calculations take much longer than docking calculations, and, in turn, very few codes support such methods. The obvious compromise would be to use a QM/MM method to describe the crucial bond formation with a quantum mechanical method while continuing to use molecular mechanics for the rest of the system. There are no codes available that do exactly this, although some that can be adapted. The QPLD (Quantum Polarized Ligand Docking) method from Schrödinger, Inc. uses a quantum mechanical calculation, but only to set partial charges, thus improving the accuracy of the electrostatic component of a molecular mechanics-based docking simulation.

Some researchers have performed docking calculations with QM or QM/MM methods by manually placing the ligand in various starting locations and doing energy minimizations or Car–Parrinello molecular dynamics. The predicted docking orientations from a conventional docking program serve as a good starting point for this type of analysis. A number of software packages have QM/MM options, such as Gaussian and the CHARMM/GAMESS combination. Some researchers have generated structures using a conventional docking program, and have then obtained the binding energies with single-point semiempirical calculations.

The only QM docking code that is available on the open market is the QUANTUM docking program from Quantum Pharmaceuticals. Other companies have kept their technologies for in-house use, such as the QM/MM docking code developed by Exegenics and the quantum mechanical free energy perturbation technique developed at Metabasis.

### 12.2.5 Hierarchical Docking Algorithms

Some vendors offer a hierarchy of docking algorithms. This is sometimes referred to as a tiered approach. One algorithm may intentionally sacrifice accuracy in order to obtain a speed sufficient for analysis of millions of compounds. Other algorithms may give more accurate results, at the expense of taking longer to run. An example of a docking program with multiple algorithms is the Glide package, sold by Schrödinger Inc.

Even when the docking package being used does not have multiple options like this, researchers sometimes set up a similar system manually. The first pass might be as simple as discarding any structure with an unreasonably large, or unreasonably small, number of atoms. The second pass might use a loose criterion based on QSAR, 3D-QSAR, or a pharmacophore search. The remaining compounds can then be run through the docking software. Figure 12.2 shows the type of steps typically incorporated into hierarchical docking processes.

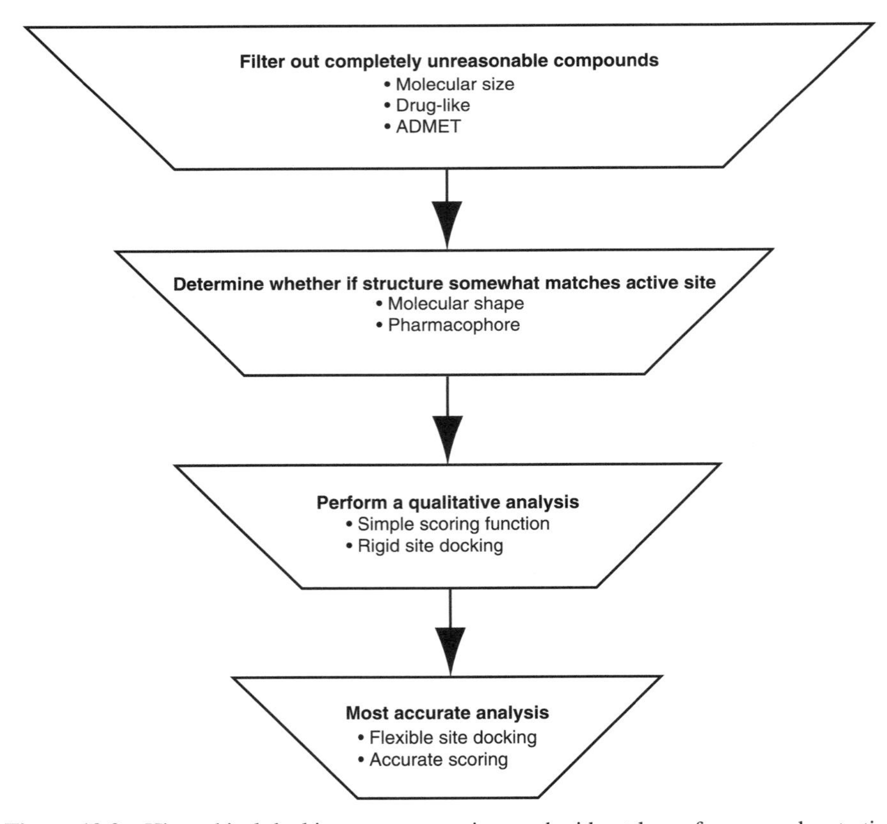


**Figure 12.2** Hierarchical docking processes gain speed without loss of accuracy by starting with steps that can eliminate compounds that clearly will not bind to the active site.

## 12.3 SCORING

### 12.3.1 Energy Expressions and Consensus Scoring

The scoring method is the single most important aspect of a docking program, as it is used for compound selection. In general, docking programs that have accurate scoring functions tend to give ligand positions that accurately reproduce results from crystallographic analysis. There are a number of different types of scoring functions, which were developed for various reasons.

Some of the simplest scoring functions are simply a quantification of how well the ligand fits in the active site. These shape-based functions generally execute quickly, but are among the least accurate scoring functions.

Most scoring functions are based on the molecular mechanics energy equation. Some are formulated as binding enthalpies or Gibbs free energies, or from a potential of mean force equation. The most accurate include entropy and solvation terms.

Consensus scoring utilizes the results of several scoring functions. The compounds that rank the best are those that were chosen by multiple scoring functions. This is an empirical attempt to improve accuracy by combining multiple scoring functions in the hope that the strengths of one will negate the weaknesses of another. At present, some of the most accurate results come from consensus scoring. Table 12.1 gives a listing of scoring functions with an explanation of the terms that they incorporate.

### 12.3.2 Binding Free Energies

In an ideal world, docking scores would be free energies of binding. In the past, most scoring functions have been energies that do not include an entropy term. Entropy may make only a small contribution to the free energy relative to the enthalpy but it becomes important when comparing more and less flexible ligands. Entropy corrections using a potential of mean force or free energy perturbation formulation are possible, but are very time-consuming. Most recently, several docking programs such as MOE and AutoDock have included a conformational entropy correction based on the torsions of the rotatable bonds in the molecule.

Many scoring functions are parameterized from sources such as experimental binding energies. Thus, the parameterization has effectively added in an entropy contribution in the energy of the single bond functions, even though the terms in the scoring equation may look like enthalpy terms only.

There are docking programs, such as the Liaison module from Schrödinger Inc., that can compute an entropy correction. This can be done as a

**TABLE 12.1 Description of Scoring Functions, which are Often Integrated with the Docking Algorithm[a]**

| Function | Description |
|---|---|
| 2D MACCS | Fingerprint similarity search |
| ADAM | Force field |
| AMBER | Force field |
| Affinity dG | (in MOE) Force field with entropy and metal binding |
| Alpha HB | (in MOE) Geometric fit plus a hydrogen bond term |
| ArgusLab | Grid-based energy |
| ASP | Atom–atom potential, consensus |
| AutoDock | GRID from force field |
| Bacon & Moult | Geometry + steric + $e^-$ |
| BLEEP | Knowledge-based |
| Catalyst | Pharmacophore |
| CHARMM | Force field |
| ChemScore | Lipophilic, metal–ligand, hydrogen bond, ligand flexibility, free energy |
| CLIX | Steric + interaction |
| ConsDock | Consensus of Dock, FlexX, and Gold |
| CScore | Consensus of D-score, ChemScore, PMF, G-score, and FlexX |
| Desjarlais et al. | Force field |
| Directed DOCK | Force field |
| Divali | Force field |
| DOCK | Shape or force field |
| DockIt | Molecular mechanics |
| DockScore | Van der Waals, electrostatic, ligand internal |
| DOCKVISION | MMFF94 or Research force field |
| DrugScore | Knowledge-based |
| D-score | DOCK-like function; electrostatic and hydrophobic |
| eHiTS | Statistically derived empirical function, optimized neural net filter |
| FEP-MD | Finite energy perturbation, molecular dynamics |
| Fischer et al. | Surface contact |
| Flex, Rarey et al. | Spherical interaction surfaces, FlexS + FlexX |
| FlexS | Alignment engine |
| FlexX (F-Score) | Docking empirical function |
| FlexX-Screen | Docking preprocesses screening |
| FlexX-Pharm | Pharmacophore based docking |
| FLOG | Hydrogen bond + hydrophobic + $e^-$ |
| FRED | Gaussian or empirical scoring |
| Fresno | Empirical |
| GB/SA | Generalized Born, solvent-accessible |
| Gehlhaar et al. | Empirical |
| Glide | Hierarchical filters |
| GlideScore | Derivative of ChemScore |
| GoldScore | Force field, empirical, same as GOLD |
| G-score | GOLD-like hydrogen bonding |
| HADDOCK | Force field |
| Hammerhead | Empirical |

(*Continued*)

**TABLE 12.1** *Continued*

| Function | Description |
|---|---|
| Hart & Read | Steric + empirical *E* |
| HINT | Empirical |
| ICM | Force field + empirical |
| Jones et al. | Hydrogen bond, van der Waals |
| Judson et al. | Force field |
| LIE | Linear interaction energy |
| LigandScore | Consensus from Ludi, Jain, PMF, PLP, LigScore |
| LIGIN | Steric + interaction + hydrogen bond |
| LigScore | Empirical, van der Waals, charges, buried charge |
| London dG | (in MOE) Force field with entropy and solvation terms |
| LUDI | Empirical/force field |
| MM2X | Force field |
| MMFF | Force field |
| M-score | Knowledge-based |
| MVP | Force field |
| NBTI | Non-Boltzmann thermodynamic integration |
| Opls-AA | Force field |
| OWFEG | Molecular dynamics |
| PBE | Pc binding elements |
| PLP (LigandFit) | Empirical, soft potential |
| PMF | Knowledge-based, interaction free energies |
| PMF-score | Statistical ligand–receptor atom-pair interaction potentials |
| PSI-Dock | Empirical |
| QPLD | Quantum charges, force field |
| QXP, Flo | Force field |
| RankScore | Empirical/force field |
| ROCS | Shape based similarity |
| SAFE_p | Empirical/force field |
| ScreenScore | Empirical/consensus |
| SIE | Empirical/force field |
| SOFT DOCK | Interaction |
| Slide | Empirical |
| SMoG | Knowledge-based |
| Surflex | Hammerhead-like, empirical, hydrophobic/polar |
| VALIDATE | Empirical/force field |
| Wallqvist & Covell | Surface burial |
| X-Score | Empirical/consensus, same as X-Cscore |

[a]Including a few pharmacophore models and similarity searches that have been compared with docking for enrichment.

post-docking calculation that incorporates additional optimization and utilizes a free energy perturbation (FEP) technique. Also, the London dG scoring function included in the MOE program from Chemical Computing Group has an entropy correction, as does version 4 of AutoDock.

### 12.3.3 Solvation

Some docking programs have the ability to include a solvation term in the scoring function energy equation. This can improve the accuracy of results. However, users should investigate the scoring functions that are available. If solvation effects were included in the data used for the parameterization of the scoring function, then an additional solvation term should not be added.

Solvation functions should add an energy correction for the solvent completely surrounding the isolated compound in solution, but only to the solvent-exposed surface (if there is any) of a compound in the protein's active site. It is not appropriate to apply a solvation term to those atoms that are in the active site and not exposed to solvent. Most solvation corrections treat the solvent as a continuum medium with a given dielectric constant. The dielectric constant is near 78 in water, but can be as low as 2–5 within the protein core.

### 12.3.4 Ligands Covalently Bound to the Active Site

Prediction of binding energies for ligands that bind covalently to the active site is problematic. Many docking codes do not allow such attachments.

Most docking codes that do allow a covalent bond between the ligand and an active site residue do so as a geometric constraint only. This means that the bond formation energy for the covalent bond is not included in the scoring energy. This can cause a significant error, since covalent bond energies are 10 times larger than hydrogen bond energies. Section 12.2.3 has more information on the physical constraint algorithms.

One solution to this problem is to determine covalent bond formation energies for each ligand using quantum mechanical calculations. In the simplest formulation, these energies can be added to the docking energy. This is not necessarily a perfect solution, since it may result in double-counting some angle bending or torsional terms.

### 12.3.5 Metrics for Goodness of Fit

A few docking programs report a goodness-of-fit metric. This is a number that indicates whether the ligand completely fills the active site, or if it leaves empty areas in the active site to be filled in by solvent molecules. Having a perfect fit to the active site is not necessarily ideal. If the fit is very tight, then slight differences in the protein due to genetic variation in the population could cause the drug to be ineffective for some individuals. If the fit is too loose, then the drug might inadvertently bind to structurally similar proteins, thus giving unwanted side effects.

## 12.4 VALIDATION OF RESULTS

When a docking study is begun, the choice of docking and scoring algorithms should be validated for that particular protein with ligands as similar as practical to those to be studied. Because scoring energies are used for compound selection in the drug design process, it is most important to make sure that the scoring energy is sufficiently accurate. This is done by comparing docking energies with biochemical assay results. Since most biochemical assays yield inhibition rate constants, this means making sure that the docking energy is proportional to ln $K_I$.

Even with the best docking programs in existence, there will not be a perfect correlation between docking energy and ln $K_I$. However, this is the most valid comparison between docking and experimental results, making it the best criterion for suggesting which docking program is best for a particular study. The correlation coefficient and graph from a validation study are a good indication of how much, or how little, accuracy to expect from the docking simulation. The docking experiments will rank compounds by predicted activity, but, as indicated by correlation coefficients, it is more correct to start with the docking energy and bin out the top 50 compounds, the next 50, and so on.

Some docking programs have scoring functions that the manufacturer claims are not supposed to correlate to experimental results. These functions may be described as qualitative results, high throughput results, or categorization methods. Under certain circumstances, these can be useful techniques—usually when they result in much faster calculations. However, researchers should be aware of the limitations of these methods when interpreting the results of computational simulations. It should be noted that there are more accurate methods on the market.

The geometry of the ligand binding conformation can also be compared with experimental results. This is done by comparing with crystallographic data. Often, a root mean square deviation (RMSD) between the computational and experimental results is presented. Unless a method gives a glaringly bad RMSD, researchers are encouraged to use this as only a very small factor in choosing a docking code. In general, methods that give accurate energies also give accurate geometries.

The amount of "enrichment" is used as an indication of the value of virtual screening (i.e., docking or pharmacophore searching), as opposed to screening all compounds with laboratory assays. For example, a large set of compounds may be screened experimentally to find that 0.1% are within some activity threshold. If a docking study is performed to choose which compounds to screen, and the subsequent screen finds 10% of this smaller set to be active, then the virtual screening is said to have given a 100-fold enrichment in results. The amount of enrichment seen with various virtual screening techniques is highly variable, ranging from 2-fold to 2000-fold enrichment. The results are generally better when the researcher doing the virtual screening is a computational chemist, who has been trained to understand the algorithmic implications of all of the software input parameters. Since computer modeling is far less expensive per compound than laboratory assays, it is economically advantageous to use virtual screening techniques to select compounds for synthesis or purchase and for laboratory screening.

## 12.5 COMPARISON OF EXISTING SEARCH AND SCORING METHODS

This section will first mention a number of the most heavily used docking programs, and then give some accuracy comparison information. The paragraphs discussing features of various docking programs mention a few noteworthy items, but are in no means a comprehensive listing of features. Most of this book does not go into even this much depth in discussing specific software packages, but, as docking is an extremely important function, it seems worthwhile to do so in this chapter. However, the reader should be aware that because of the appearance of new software offerings and improvements in functionality of existing codes, some of the detailed information presented here may become out of date fairly rapidly.

The first docking program, DOCK, was created by Irwin Kuntz at the University of California, San Francisco (http://dock.compbio.ucsf.edu/). Now at version 6, this program has evolved over the years. It originally docked a rigid molecule to a rigid active site using a geometric fit. Today, it allows molecules to be flexible and uses force field-based scoring. Most recently, the program has been enhanced to allow the active site to be flexible and includes a solvation term in the scoring function. It ships as source code, which is primarily C++, with some utility programs written in Fortran. The scoring options include Delphi electrostatics, ligand conformational entropy corrections, ligand desolvation, receptor desolvation, GB/SA solvation scoring, PB/SA solvation scoring, and AMBER scoring (including receptor flexibility and the full AMBER molecular mechanics scoring function with implicit solvent). It also uses conjugate gradient minimization, and has molecular dynamics simulation capabilities.

AutoDock is a suite of docking tools developed at The Scripps Research Institute, (http://autodock.scripps.edu/). AutoDock 4.0 is currently in beta testing. It allows side chains in the macromolecule to be flexible, as well as allowing for ligand flexibility. AutoDock 4.0 adds a free-energy scoring function created from a linear regression analysis, the AMBER force field, and a large set of diverse protein–ligand complexes with known inhibition constants. Previous versions of AutoDock used a scoring function that computed a binding energy based on the AMBER force field, but was reparameterized specifically for docking. The searching algorithms include simulated annealing and a genetic algorithm. The accuracy comparisons below are based on AutoDock 3 and earlier.

Accelrys, presently the world's largest chemistry software company, sells the LigandFit docking module, along with the LigandScore scoring module (http://accelrys.com/). These can be used with either the Cerius$^2$ or Discovery Studio graphical interfaces from Accelrys. LigandFit includes functions for identifying the active site of the receptor. This is valuable for very large scale processing of data, in which every molecule needs to be docked into every receptor with no human intervention. Two scoring functions are included in LigandFit: PLP (piecewise linear potential), which gives a smoothed energy landscape, and the DockScore potential energy function. An energy grid for faster docking can be created using either the Dreiding or CFF force fields. The optional LigandScore module adds in the LigScore empirical scoring functions, the PMF (potential of mean force) knowledge-based scoring function, Jain (an empirical scoring function that includes entropic and solvation effects), the Ludi Energy Estimate methods, and a consensus scoring option.

The docking program FlexX is sold by BioSolveIT. At one time, FlexX was resold by Tripos for use within the SYBYL graphical interface. This is one of the older docking codes on the market. The unique feature of FlexX is a molecule positioning algorithm that breaks the molecule into fragments, and then builds it within the active site. The traditional strengths of FlexX have been a rather fast docking algorithm and the ability to give a reasonable placement of the ligand in the active site. Critics of FlexX tend to cite a less than desired accuracy of binding energies. When FlexX was sold by Tripos, researchers tended to use FlexX in conjunction with the CScore consensus scoring module from Tripos. Researchers in the pharmaceutical industry tend to use FlexX as an active site shape fitting tool, but seldom for quantitative prediction of drug activity.

Tripos, currently the world's second largest chemistry software company, has stopped selling FlexX and has replaced it with Surflex-Dock. Surflex-Dock was developed by Ajay Jain at the University of California, San Francisco. The docking algorithm is a flexible alignment algorithm that uses a crossover procedure to generate new poses by combining pieces of intact, aligned molecules. The scoring function is an energy equation that has been parameterized from known binding affinity data. The parameterization included negative training data to reduce false-positive results. Tripos also sells CScore, which is a consensus scoring program designed to identify molecules that are shown to bind well by five different scoring algorithms: D-score, ChemScore, PMF-score, G-score, and FlexX.

Chemical Computing Group sells a docking function integrated into its MOE drug design software package. The docking program uses a tabu algorithm to carry out flexible docking of molecules in the active site, or it can score poses from a database. The binding energy, based on a London dG scoring function, can be computed using either a grid-based algorithm, or the full force field. Entropy and solvation effects can be included in the energy expression. There is also an Affinity dG algorithm that includes force field, entropy, and metal binding terms, and an Alpha HB scoring function that has only a geometric fit and a hydrogen bond term. Chemical Computing Group distributes full source code for this algorithm, written in the SVL programming language.

Glide is the docking module sold by Schrödinger, Inc. It can be accessed from the Maestro graphical interface from Schrödinger. Glide offers several unique features. It has three options for speed/accuracy of prediction: a high throughput virtual screening (HTVS) mode, standard precision (SP), and extra precision (XP). The HTVS mode can be used as a filtering step to cut down the size of a very large collection of molecules, followed by more accurate docking calculations with the other modes. Glide can be combined with the Liaison package from Schrödinger, in order to compute free energies of binding.

QXP is the docking code that comes in the Flo software package sold by Thistlesoft. Thistlesoft manages to sell some expensive source code licenses, in spite of the fact that they do not have a sales staff, or even a web page. The docking code is distinguished by having a rich set of features "under the hood." This makes it more difficult than other docking programs for the user to learn to utilize effectively, but can be an advantage when docking to chemical systems where difficulties are encountered when using other docking codes.

ICM-Dock from MolSoft is a flexible docking program that uses the MMFF force field. The force field can be used to generate a grid potential for faster docking. This is one of the few docking programs that advertises the ability to do protein–protein docking.

GOLD is a docking program developed by a collaboration between the University of Sheffield, GlaxoSmithKline, and the Cambridge Crystallographic Data Centre. GOLD is widely used in the pharmaceutical industry. It uses a fast genetic algorithm for protein–ligand docking. The algorithm can allow ligand flexibility and a measure of active site flexibility. The GoldScore, ChemScore, and ASP scoring functions come with it, and it has the ability to plug in a user-defined scoring algorithm. The Hermes graphical interface is one of the more fully featured graphical interfaces available for a standalone docking program. The GoldMine program, which comes with GOLD, is used for postprocessing docking results.

There are two DrugScore scoring functions: DrugScore$^{PDB}$ and DrugScore$^{CSD}$. These are scoring functions that do not come bundled with a specific docking code. The unique features of DrugScore are the ability to score a molecule via a web page, as well as a visualization that shows the magnitude of the contribution to the final score ascribed to each atom. It uses a knowledge-based scoring expression.

X-Score (formerly known as X-CScore) is a scoring function that can be used with DOCK, AutoDock, FlexX, or GOLD. It is distributed free, including a manual and source code. It is developed at the University of Michigan, with a website at http://sw16.im.med.umich.edu/software/xtool/.

FRED is a docking module sold by OpenEye Scientific Software. It takes a set of initial conformations, often generated by the OMEGA package. The initial poses are both docked and optimized by FRED. FRED has two steps to the docking algorithm. The first is a shape fitting step that places the molecule in the active site. The second step is optimization of the position, which is done by optimizing hydroxyl hydrogen atoms, rigid ligand optimization, and optimizing all dihedral bond angles. The scoring functions available are Gaussian shape scoring PLP, ChemScore, and ScreenScore.

There are many other docking programs. These include ADAM, DARWIN, DIVALI, DockVision, EUDOC, FLOG, eHiTS, FTDOCK, LIGIN, MCDOCK, PRO_LEADS, ProDock, SANDOCK, and Soft docking. The reader is urged to investigate any that seem suitable for their needs, and to carry out a comparison study at the beginning of their research project in order to choose the one that is best for the chemical system being studied.

A significant number of papers have been published with comparisons or benchmarks for docking program performance. However, these publications are not necessarily directly comparable with one another, because there is no universally established measure of docking performance. Publications most often use one or more of three common comparisons: (1) the ability to reproduce the ligand position identified from X-ray crystallography experiments; (2) the accuracy of predicting binding free energies in ranking the same as biochemical assays; and (3) the ability to discriminate known inhibitors from other drug-like molecules given as an enrichment. Fortunately, there seems to be a fairly high degree of correlation between these metrics. Table 12.2 gives a summary of many docking analysis studies. Since the results are not directly comparable, it lists how many times each docking program was tested and the number of times that program came out best. As such, Table 12.2 is primarily a view as to how well programs work across a wide variety of chemical studies. Some papers list several programs as being best within the accuracy of the study, which is reflected in Table 12.2. The complete list of references used to generate this table is in the supplemental references on the CD accompanying this book. Some docking programs listed in Table 12.2

**TABLE 12.2 Docking Comparison Studies[a]**

| Docking Method | No. of Times Tested | No. of Times Best |
|---|---|---|
| 2D_MACCS | 3 | 0 |
| ADAM | 1 | 1 |
| AMBER | 1 | 0 |
| AMBER + PBE | 1 | 1 |
| ArgusLab | 1 | 0 |
| AutoDock | 14 | 1 |
| Bacon and Moult | 1 | 0 |
| Bohacek | 1 | 1 |
| Born + ASP | 1 | 0 |
| Catalyst | 3 | 0 |
| CHARMM | 1 | 0 |
| ChemScore | 15 | 1 |
| ChemScore-GS | 1 | 0 |
| CLIX | 1 | 1 |
| ConsDock | 14 | 4 |
| Cscore | 9 | 3 |
| Deprecated RS | 9 | 0 |
| Desjarlais et al. | 1 | 0 |
| Directed DOCK | 1 | 1 |
| Divali | 1 | 0 |
| DOCK | 48 | 5 |
| DockIt | 8 | 1 |
| DOCKVISION | 8 | 0 |
| DrugScore | 17 | 4 |
| D-Score | 20 | 2 |
| eHiTS | 1 | 2 |
| FEP-MD | 5 | 4 |
| Fischer et al. | 1 | 1 |
| Flex, Rarey et al. | 1 | 0 |
| FlexX (F-Score) | 69 | 18 |
| FlexX-Screen | 1 | 1 |
| FLOG | 2 | 0 |
| FRED | 19 | 2 |
| GB/SA | 1 | 0 |
| Gehlhaar et al. | 1 | 0 |
| Glide | 34 | 3 |
| GoldScore | 56 | 15 |
| Grid | 1 | 0 |
| G-Score | 20 | 3 |
| Hammerhead | 1 | 0 |
| Hart and Read | 1 | 1 |
| ICM | 12 | 5 |
| Jones et al. | 1 | 0 |
| Judson et al. | 1 | 0 |
| Kasper | 1 | 1 |

(*Continued*)

**TABLE 12.2** *Continued*

| Docking Method | No. of Times Tested | No. of Times Best |
|---|---|---|
| LIE | 2 | 2 |
| LigScore | 11 | 5 |
| LIGIN | 1 | 1 |
| LUDI | 15 | 0 |
| ME | 4 | 1 |
| MM2X | 1 | 0 |
| MOE (obsolete version) | 8 | 0 |
| MVP | 8 | 4 |
| NBTI | 1 | 1 |
| OWFEG | 1 | 0 |
| PBE + ASP | 4 | 1 |
| PLP | 33 | 14 |
| PLP1 | 1 | 1 |
| PLP2 | 1 | 1 |
| PMF | 28 | 2 |
| PMF-Score | 2 | 1 |
| QXP, Flo | 17 | 2 |
| Rank-sum | 9 | 0 |
| ROCS (enrichment only) | 11 | 9 |
| Rognan | 1 | 1 |
| Slide | 8 | 0 |
| SOFT DOCK | 1 | 0 |
| Surflex | 10 | 7 |
| Thermodynamic Integration | 1 | 1 |
| VALIDATE | 1 | 0 |
| Viswanadhan | 1 | 1 |
| Wallqvist and Covell | 2 | 1 |
| Worst–best | 9 | 0 |
| X-Score | 10 | 3 |

[a]Studies may have been ranked based on RMSD compared with crystal structure or on enrichment. Some studies compared docking with pharmacophore methods for enrichment, so the pharmacophore results are included as well.

may have done better or worse than the reader may have expected. Here are some points to note about these results:

- Some of the docking programs that have been around the longest and used the most did more poorly than expected. This may be because programs made later included improvements created based on the lessons learned from those earlier efforts.
- Some are listed in very few studies. Some of these were published, but not incorporated into publicly available programs. Others are so new that a large body of literature utilizing them has not yet been published.

- The MOE program from Chemical Computing Group was included in only one study that did a number of tests, in spite of the fact that it is used widely in the industry. The docking and scoring functions have been significantly rewritten since this study was carried out. This is a "best of breed" docking program that scores using a force field with entropy and solvation corrections.
- Many of these studies tested docking programs against just one or a few targets. Very few studies tested against a very wide selection of targets. Table 12.2 lists numbers of studies rather than numbers of targets—thus suggesting the potential for smaller studies to skew the results somewhat.

The most frequently cited docking software is AutoDock. However, the results in Table 12.2 suggest that AutoDock is not the most accurate docking program. This metric is skewed by the fact that AutoDock is free for academic use, whereas the majority of docking packages are sold commercially. Since academic researchers publish most prolifically, whereas industry researchers patent most prolifically, we can surmise that AutoDock is the docking software most frequently used in academic settings. It should also be noted that none of these studies tested the recently released version 4 of AutoDock.

There have been many docking comparison studies, but few very comprehensive ones. Some compared results only with the least accurate of the competing techniques. Others examined a very small number of chemical systems, possibly unconsciously chosen based on the strengths of the author's docking code. The following paragraphs point out a couple of the docking comparison studies that are the most impressive in their size and scope.

Of the docking accuracy comparison studies, one of the largest is the study by Kellenberger, Rodrigo, Muller, and Rognan. This study is particularly impressive because it compared results with 100 protein–ligand complexes from the Protein Data Bank (PDB). The docking programs tested were DOCK, FlexX, FRED, Glide, GOLD, Slide, Surflex, and QXP. In this study, Glide, GOLD, and QXP did best at reproducing the docked geometry from the PDB. Glide, GOLD, and Surflex did the best at ranking compounds. Surflex and GOLD did the best at having known actives fall in the top 10% of the rankings of a set of test compounds for a thymidine kinase example.

Another large study was carried out by Wang, Lu, and Wang. They found that PLP, F-Score, LigScore, and DrugScore did well on many systems. X-Score did best for predicting poses of hydrophilic ligands, but poorly overall on geometric predictions. Conversely, X-Score did best for reproducing binding affinities, when given the correct ligand conformation. They performed various tests using multiple scoring functions to create a consensus scoring scheme. The best consensus of two scoring functions came from DrugScore plus LigScore, but many other combinations of two scoring functions did nearly as well. Two combinations of three scoring functions came out well: LigScore plus DrugScore plus F-Score and LigScore plus DrugScore plus PLP. A number of other triple scoring combinations did nearly as well.

Note that the above discussion and most of the literature results are based on a typical scenario of docking an organic molecule to a protein's active site. This makes it even more important to carry out a careful evaluation of the method when docking small molecules to active sites containing metals or when simulating inorganic compounds. Also, docking between two protein units or between protein and RNA are outside the scope of these studies.

The metrics given above are an overall indication of docking performance. Not surprisingly, some programs may do better or worse, depending upon the natures of the ligands and active site. There are a few studies that give some indication as to how docking programs perform depending upon whether ligands are large, small, polar, hydrophobic, flexible, rigid, etc. and whether the active site is small, open, etc. One such comparison is given in Table VII of the Kellenberger et al. paper referenced at the end of this chapter. Another is in Wang et al.

## 12.6 SPECIAL SYSTEMS

The previous section has given some comparisons of specific docking algorithms and scoring functions. These comparisons tend to be reasonable metrics if docking is being used for the types of chemical systems that it was designed to model. The applicability of docking can be extended to cover other types of chemical systems, but accuracy comparisons from previous studies are generally not a good measure of how well the docking will work in these cases. This section discusses some systems for which docking can be used, but with greater caution and careful validation.

Docking programs are designed to predict whether a ligand will inhibit the action of an enzyme through a competitive inhibition mechanism. This statement implies several limitations of docking. Not all drugs function through competitive inhibition. Not all biomolecules are proteins. As such, there are systems that are stretching the applicability of docking algorithms, and systems where docking is not applicable. Since validation studies are performed on ligands that operate through a competitive inhibition mechanism, researchers should always carry out an independent validation for the system under investigation if it is not a competitive inhibition system.

Some drugs bind at allosteric sites on enzymes. Most allosteric interactions function as control or feedback mechanisms in the metabolic pathway, thus making this a way to obtain a very potent drug. In some cases, it is possible that the drug is still functioning through a competitive inhibition mechanism—just inhibiting the binding of the moderator compound rather than the active site. However, often drugs that bind at allosteric sites function as moderators with enhanced or decreased potency. Since competitive inhibitors at the active site nearly always downregulate a metabolic pathway, having drugs that are potent moderators bind at the allosteric site is one way of upregulating the metabolic pathway.

Having a drug that is a moderator binding to an allosteric site creates significant difficulties in predicting drug activity. Binding to the allosteric site is necessary, but often not sufficient to result in strong drug activity. The researcher must examine the mechanism by which the native moderator influences the protein's function. The moderator may alter the protein's shape or vibrational modes. It may also be an electron or proton donor or acceptor. Once this mechanism is understood, the researcher must devise a way to compute a quantitative measure of the potency of each ligand with regard to this mechanism. Mechanisms that involve electron or proton transfer may be predicted using quantum mechanical calculations. Mechanisms that modify the protein's shape or vibrational behavior often require the use of optimization, molecular dynamics, or molecular docking calculations, combined with some mechanism for measuring the crucial distance, shape, etc.

Docking to DNA entails some technical difficulties. Some docking programs work best when there is a well-defined active site cavity or crevice in a protein. However, some docking programs will perform well for binding in the shallower DNA grooves. Drugs that bind by intercalation of DNA base pairs are often modeled poorly by docking, even with flexible active site algorithms. Intercalated compounds are often better modeled using Monte Carlo or molecular dynamics software.

The active conformations of RNA strands look similar to enzymes, with clefts and pockets. Docking can be used to simulate the inhibition of RNA by small molecules (as opposed to antisense compounds). However, there are some caveats that should be considered. RNA tends to be more flexible than a folded protein, thus making active site flexibility very important. Scoring functions that compute an intermolecular binding energy generally work reasonably well for RNA–ligand binding. The more empirical scoring functions generally do poorly for modeling interactions with RNA.

When carrying out docking studies of steroids, the scoring function should be validated on a steroid test set to ensure that it gives good results for the lipophilic steroids. The other important precaution to take with steroid docking is that the active site may undergo two different conformational changes, depending upon which steroid binds to it. It is usually necessary to check manually which change is occurring, usually by measuring an interatomic distance that is indicative of the possible outcomes.

## 12.7 THE DOCKING PROCESS

The following subsections discuss the process of setting up and running a docking calculation. This is illustrated graphically in Fig. 12.3.

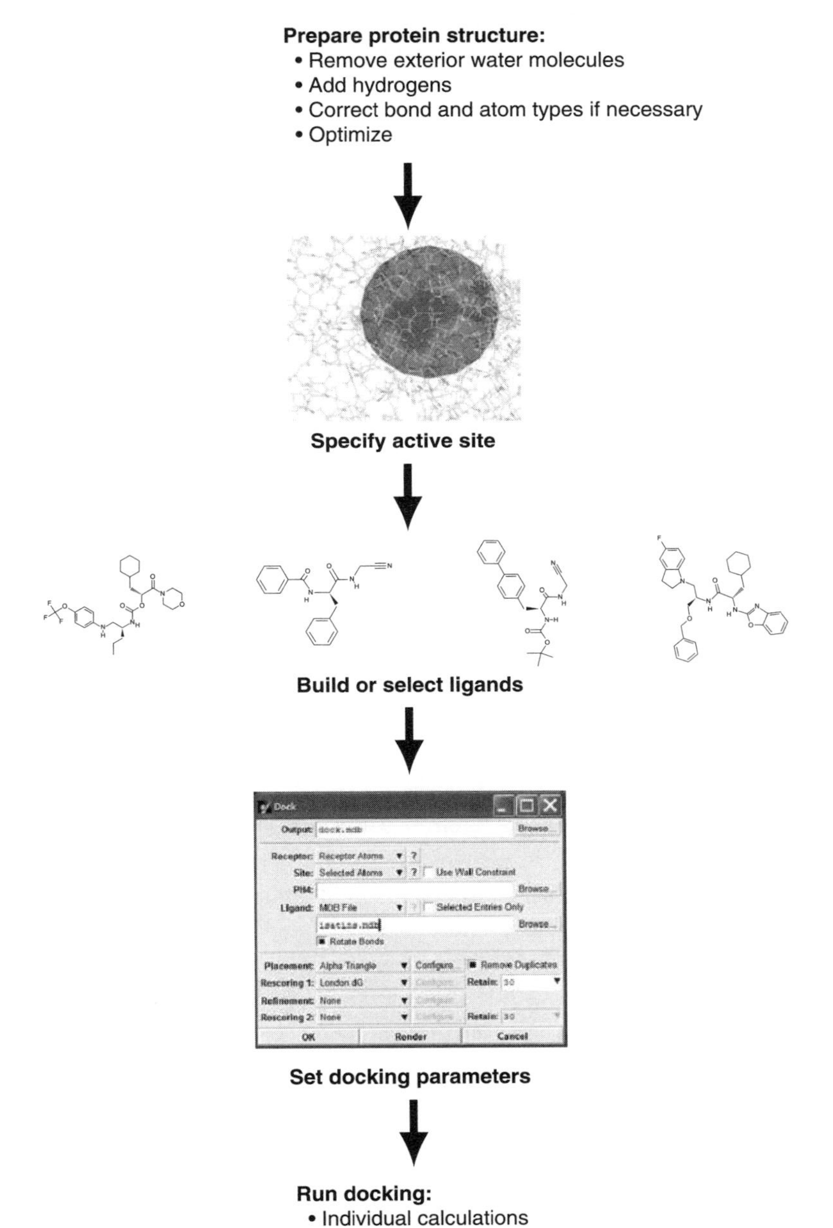


**Figure 12.3** Illustration of the docking process.

### 12.7.1 Protein Preparation

The accuracy of docking results is directly related to the quality of the crystallographic structure for the protein's active site. If the enzyme requires a cofactor, then the crystallographic structure of the holoenzyme (with the cofactor attached) should be used. Since even the best crystallographic structures often have a resolution of the order of 1 Å or more, most researchers will start with the crystallographic structure with an inhibitor in the active site, and then add the hydrogens that cannot usually be seen by protein crystallography. Water molecules are removed, sometimes with the exception of structural waters in the interior of the protein. Finally, a simple minimization is performed to find the nearest minimized structure, as predicted by the force field being used. This minimized structure is often the best structure to utilize for docking studies, particularly if the crystallographic resolution was marginal.

### 12.7.2 Building the Ligand

Docking programs offer several options for creating ligands and placing them in the active site. At the most automated end of the spectrum, some docking programs can take a database of ligands, place each one in the active site, and do the docking run. This completely automated approach allows thousands of compounds to be analyzed without manual intervention from the user. Some docking programs are slightly less automated in that they automatically dock a list of ligands from a database, but each ligand must be stored in the database with its position, orientation, and conformation set within the coordinate system to put the initial placement in the protein's active site.

A variation on the automated approach is to first generate all of the ligand poses (position, orientation, and conformation) and store them in a file. These poses can then be automatically read in and a docking score computed without any automated movement of the ligand. This is as close to ideal as possible for comparing scoring functions, independently of the search algorithm. It is not a perfect test of how well different docking programs will work, because some search algorithms give good results because they are using the scoring results to guide the searching. Storing all of the poses can take up significant disk space, and researchers seldom have any reason to examine more than a handful of the poses with the best scores. Because of this, most docking programs take a single pose as an input, generate many thousands of poses, but only store the best ones (perhaps 20–100, as chosen by the user) as the final results.

At the opposite end of the spectrum from the highly automated docking programs are programs that require the user to place the ligand manually in the active site, and then run each docking simulation separately. This is not necessarily an inconvenient way to perform docking calculations. Much structure-based drug design work is done by manually building compounds in the active site of the target protein. The drug design chemist will work with a graphical interface that can display the 3D shape of the protein and the ligand being built on a desktop computer. Many of these interfaces have features to aid in this visual examination, such as putting in a dotted line to indicate when two atoms are within a reasonable distance and orientation to form a hydrogen bond. The drug designer can build up chemical structures in the active site by choosing functional groups that will fit in the space, form hydrogen bonds to the protein, etc. Sometimes, they start with the native substrate for an enzyme, and then modify the molecule so that the enzyme will not react with the new compound (e.g., by replacing a good leaving group with a group that will not dissociate).

### 12.7.3 Setting the Bounding Box

The entire protein is read into the computer in order to obtain a correct geometry for the structure. However, docking calculations can take a significant amount of time, and anything that speeds them up without loss of accuracy is utilized. Protein residues far from the active site do not generally have any measurable effect on the scoring results. In order to speed the calculation, a cutoff distance is set, and no interactions are computed for residues beyond that distance from the ligand. For ease of coding, this cutoff is usually a rectangular box, called a bounding box. In the case of grid potentials, the grid is only computed for points within the cutoff distance. Many docking packages default to setting the bounding box far enough out from the ligand or active site that there is no need to alter it. However (at least when starting on a set of docking runs), the user should verify that the bounding box extends a reasonable distance beyond the active site in all directions.

### 12.7.4 Docking Options

When setting up the inputs to a docking calculation, there will be options for flexible active sites, scoring method, search method, solvation, handling encapsulated active sites, etc. Some packages may have an option to do a fast initial check, and then continue only if the first check criterion is satisfied. Nearly always, the docking results will be more accurate with the calculation running more slowly, if an optimization (or at least a few optimization steps) is performed for each pose.

It is well worth the drug designer's time to read up on all the options available in a docking package. It is also highly advisable that a set of validation tests be performed at the beginning of a drug design project. If a given parameter seems to work best when set far from the manufacturer's recommendation, then additional attention should be given to determining why there is a large discrepancy. Once the best input options have been determined by the validation tests, the same settings will be used for every docking calculation done during the project.

Drug designers frequently keep old versions of the docking software active on their computer systems. A project spanning years will have every calculation done with the version used for the original validation at the beginning of the project. Thus, each project may be done with a different version of the software.

### 12.7.5 Running the Docking Calculation

Once the inputs have been set, the docking calculation can be run. Most docking programs allow individual docking runs to be started from the graphical interface. A few docking programs can run many docking calculations in a batch mode. These calculations are sometimes run on the same computer where the graphical interface is used, and sometimes can be sent off to a different server.

### 12.7.6 Analysis of Results

The most important result from a docking calculation is the binding energy of the ligand to the active site. This is the value that is compared between different compounds to determine which will be the best inhibitor. The pose of the ligand in the active site associated with a few of the best binding energies is examined visually to ensure that it looks reasonable. Sometimes, the pose generated by one docking calculation gives the researcher an idea for how to alter the compound on the next round of calculations.

The fit to the active site should be examined. Functional groups that hang out of the active site may be extraneous, and suggest that having a solvation term in the scoring function might be particularly important for this target. Areas where the molecule does not fill the active site indicate spots where a functional group could be added to give stronger binding. Too tight a fit into the active site, particularly of a pathogen protein, may indicate that resistance to the drug could be built up easily by a minor modification of one of the active site residues.

There are rare cases where the lowest energy binding orientation is not the biologically active binding mode. When this happens, the biologically active pose must be within 10 kcal/mol of the lowest energy pose, in order to be energetically accessible at biological temperatures.

As pointed out earlier, validating a docking program for the specific target and class of ligands being designed is often the most important criterion for selecting a docking program. However, some generalities can be made about the docking programs on the market:

- Selection of the best docking program is dependent on the nature of the protein and ligands.
- Inclusion of active site flexibility is critical for some targets.
- Inclusion of solvation effects gives an improvement in the accuracy of results for most targets.
- ICM, Glide, and GOLD work well for enclosed binding sites.
- Surflex, GOLD, Glide, LigScore, and MOE are generally fast and reliable for a diverse range of binding sites.
- Of those that were included in 10 more tests in Table 12.2, ROCS, a shape-based active site fitting tool, came out best for enrichment. Of the quantitative scoring functions, LigScore, PLP, ICM, and X-Score came out best.
- Consensus scoring is sometimes better than scoring with a single function.
- Energy minimization of docked poses can give a significant improvement in results.
- Inclusion of entropy corrections can give an improvement in the accuracy of results when comparing more and less flexible ligands.

Docking is the primary workhorse of structure-based drug design. As such, a detailed knowledge of the software tools being used is valuable to a drug designer.

## BIBLIOGRAPHY

### Books

Alvarez J, Shoichet B. Virtual Screening in Drug Discovery. Boca Raton, FL: CRC Press; 2005.

Huang Z. Drug Discovery Research: New Frontiers in the Post-Genomic Era. Hoboken, NJ: Wiley-Interscience; 2007.

Stroud RM, Finer-Moore J, editors. Computational and Structural Approaches to Drug Discovery: Ligand–Protein Interactions. London: Royal Society of Chemistry; 2007.

## Review Articles and Accuracy Comparisons

Note that the complete set of references used to generate Table 12.2 is in the supplemental references file on the accompanying CD.

Böhm HJ, Stahl M. The use of scoring functions in drug discovery applications. Rev Comput Chem 2002; 18: 41–87.

Kellenberger E, Ridrigo J, Muller P, Rognan D. Comparative evaluation of eight docking tools for docking and virtual screening accuracy. Proteins 2004; 57: 225–242.

Muegge I, Rarey M. Small molecule docking and scoring. Rev Comput Chem 2001; 17: 1–60.

Oshiro CM, Kuntz ID, Knegtel RMA. Molecular docking and structure-based design. In: von Ragué Schleyer P et al., editors. Encyclopedia of Computational Chemistry. New York: Wiley; 1998. p. 1606–1613.

Sousa SF, Fernandes PA, Ramos MJ. Protein–ligand docking: Current status and future challenges. Proteins 2006; 65: 15–26.

Wang R, Lu Y, Wang S. Comparative evaluation of 11 scoring functions for molecular docking. J Med Chem 2003; 46: 2287–2303.

Additional references are contained on the accompanying CD.

# 13

# PHARMACOPHORE MODELS

When researchers start on a drug design project, one of the early steps is to assay available compounds in order to obtain some initial active hits. Of course, it is prohibitively costly to simply assay every compound available. Thus, there needs to be a logical process for choosing compounds with some reasonable probability that any one of them might be a useful inhibitor. The tools that fill this need are substructure searches, similarity metrics, and pharmacophore searches. Of the three, pharmacophore searches are best at finding a range of chemical structures with viable features, and are thus the method of choice for the first round of compound selection. This ability of a pharmacophore model to find new classes of inhibitors when one class is known is called "scaffold hopping."

Once a few active compounds have been found, it is very easy to run substructure searches of large compound databases. Substructure searches find any compound that contains the exact pattern of atoms and bond given in the search query. For example, a substructure search can find all compounds containing an identical fused ring system. Substructure searches are certainly a useful starting point. However, they fail to find very similar compounds, such as analogs that have five-membered rings replaced by six-membered rings.

Similarity searching goes one step further in that it can find close analogs. Similarity searches can find compounds with a slightly different ring systems, ring-opened analogs, etc. However, similarity searches also have their

*Computational Drug Design.* By David C. Young

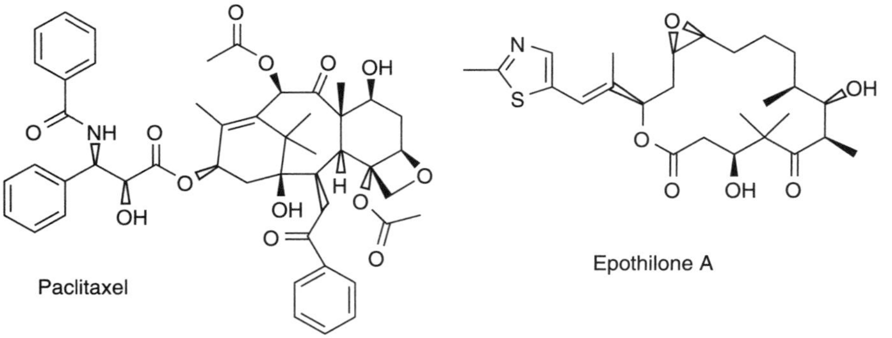

**Figure 13.1** Taxanes (e.g., paclitaxel) and epothilones both bind to the tubulin binding site, in spite of having very different chemical structures.

limitations. It is possible to have several compounds that have extremely different backbones but all bind to the same active site. For example, taxanes and epothilones (shown in Fig. 13.1) both bind in the tubulin active site. Similarity and substructure searching is described in more depth in Chapter 18.

In the taxane/epothilone example, the reason that both compounds are active is not because they share a structural similarity, but rather because they both fit the same pharmacophore. A pharmacophore is a description of the properties needed for a compound to bind in an active site. A pharmacophore describes a three-dimensional arrangement of molecular features: hydrogen bond donors and acceptors, bulky groups, etc. A pharmacophore search program is one that searches through large collections of molecules to find those that contain a pharmacophore. Figure 13.2 shows a taxane and an epothilone against the same pharmacophore.

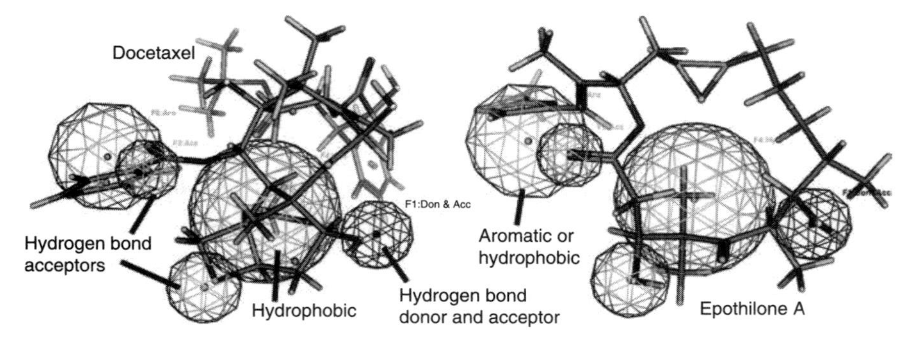


**Figure 13.2** A taxane (docetaxel) and an epothilone (epothilone A) against the tubulin pharmacophore (aligned from pdbs 1TUB and 1TVK).

## 13.1 COMPONENTS OF A PHARMACOPHORE MODEL

A pharmacophore consists of a three-dimensional arrangement of chemical features. The list of features supported differs from one software package to another. Milne presents pharmacophores in terms of enzyme-ligand interaction types as follows:

- ion–ion
- ion–dipole
- dipole–dipole
- hydrogen bonds
- induced polarization
- charge transfer
- dispersion forces
- steric repulsion

Some pharmacophore searches, particularly those from earlier in the development of pharmacophore searching, will search for a particular arrangement of specific atoms: for example, a carbonyl, a benzene ring, and a hydroxy group at specified distances apart. Most pharmacophore descriptions in the more recently developed software packages are more general. Some features typically available for inclusion in these pharmacophores are:

- hydrogen bond donors
- hydrogen bond acceptors
- aromatic rings (may be ring atoms, ring center, or normal to the ring)
- hydrophobic centers (also called neutral centers)
- positive charge centers
- negative charge centers
- acidic groups
- basic groups
- bulky groups engaged in steric interactions
- planar atoms
- $CO_2$ centroid (i.e., ester or carboxylic acid)
- $NCN^+$ centroid
- metal (also called a metal ligator)
- excluded volumes – forbidden regions, where the protein is and the ligand cannot have functional groups

Some programs can impose additional constraints on the centers, such as noting that one of the above centers is only satisfied if the atom is in a chain or in a ring.

Some programs can produce an $IC_{50}$ estimate based on a curve fit to the pharmacophore model. These tend to be very crude estimates, sometimes in error by several orders of magnitude. The strength of pharmacophore searching is in identifying compounds that should be tested for activity, not in quantitative prediction.

## 13.2 CREATING A PHARMACOPHORE MODEL FROM ACTIVE COMPOUNDS

One of the advantages of pharmacophore searching is that it is possible to generate a pharmacophore model without any direct knowledge of the geometry of the protein's active site. This can be done by analyzing a set of known active ligands to find what features they have in common. Figure 13.3 shows an example of a pharmacophore being created from active compounds.

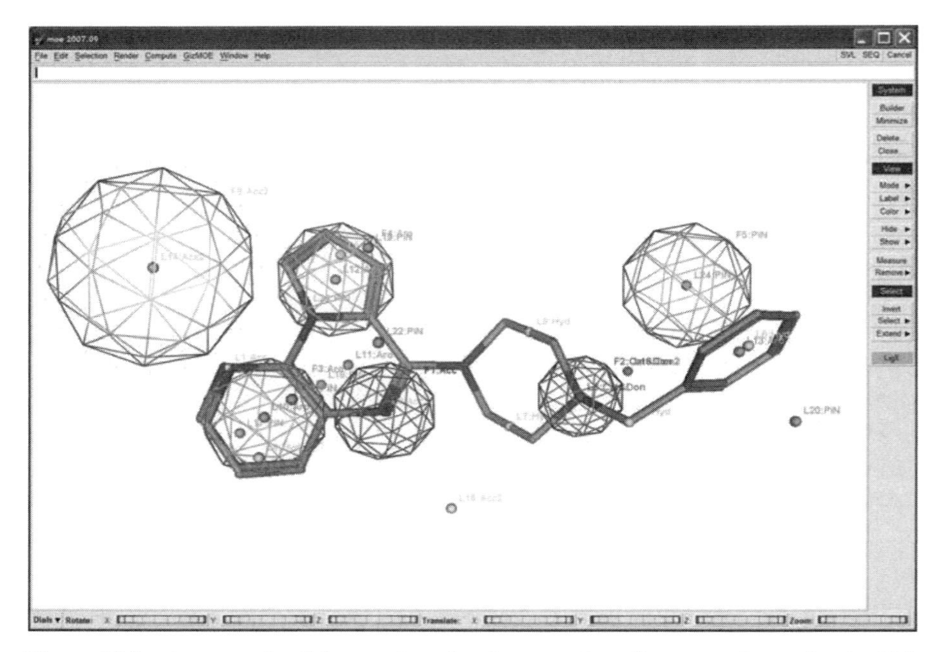


**Figure 13.3** An example of the creation of a pharmacophore from an active molecule. This example shows the MOE software from Chemical Computing Group, which indicates possible positions for pharmacophore features, but lets the user select and adjust those features manually.

In practice, this can be somewhat difficult. One reason for this is that there is no sure way to know what conformer of an active ligand is the active conformer. Worse yet, the active site may have an important functional group suggesting a pharmacophore feature that is not present in any of the known ligands. Also, the known ligands may by chance have a feature not required for binding, particularly if they are all derivatives of the same base structure. This is why it is best to perform this analysis with a set of active ligands that are not all derivatives of the same structure. In spite of these difficulties, constructing a pharmacophore model from known ligands may be the best option available at some early stage of the project.

The first step in building a pharmacophore model from a set of ligands is the alignment of the molecules, including position, rotation, and conformation. The initial alignment may be done with active compounds only. Later, inactive compounds can be aligned onto this set. Ideally, the alignment starts with the most rigid fused ring compounds; other compounds are then aligned to those. This is a mechanism for finding the correct conformer of each compound to use. If there are no rigid structures, the "shoot-from-the-hip" alternative is aligning the compounds to the lowest energy conformer of the most active compound.

This alignment can be done by hand or using an automated algorithm. The hand method requires the researcher to specify which features should be aligned with one another for each pair of molecules. The automated algorithms generate a list of conformers for each compound, and then try to find an alignment of best fit for the whole list. Often, this automated algorithm works very successfully. Sometimes, the algorithm can force the molecules into conformation that would not be energetically accessible in order to get a best fit across a range of compounds that may include one that is an outlier having a vital pharmacophore feature missing or in a nonoptimal location.

The algorithm to automatically generate a pharmacophore model must determine what features are in all molecules. This is called a largest common point set algorithm. In theory, this is a very computationally intensive NP-complete algorithm. However, in practice, it does not usually pose a significant problem, because the number of known ligands with reasonably strong activity tends to be fairly small. This process can be done by an automated algorithm or manually.

Many researchers prefer to have some manual control over the selection of pharmacophore model features. Some programs, such as MOE, are design to facilitate a hand-building process. These programs suggest potential features, but leave it up to the user to select which to use, and manually modify their positions. Often, the pharmacophore is first made to fit the most active compounds, and then modified to exclude the worst compounds.

A correlation coefficient can be computed as a measure of fit between predictions from pharmacophore models and test-set compound activities. This is often used to choose which of multiple possible pharmacophore models to utilize.

## 13.3 CREATING A PHARMACOPHORE MODEL FROM THE ACTIVE SITE

The alternative to generating pharmacophore models from active compounds is to generate them from the geometry of the active site. This is usually the preferred option if this geometry is known. Since it is possible to see how the pharmacophore model interacts with the active site, there will not be errors due to having a limited set of training compounds. Figure 13.4 shows an example of a program for creating a pharmacophore from the active site.

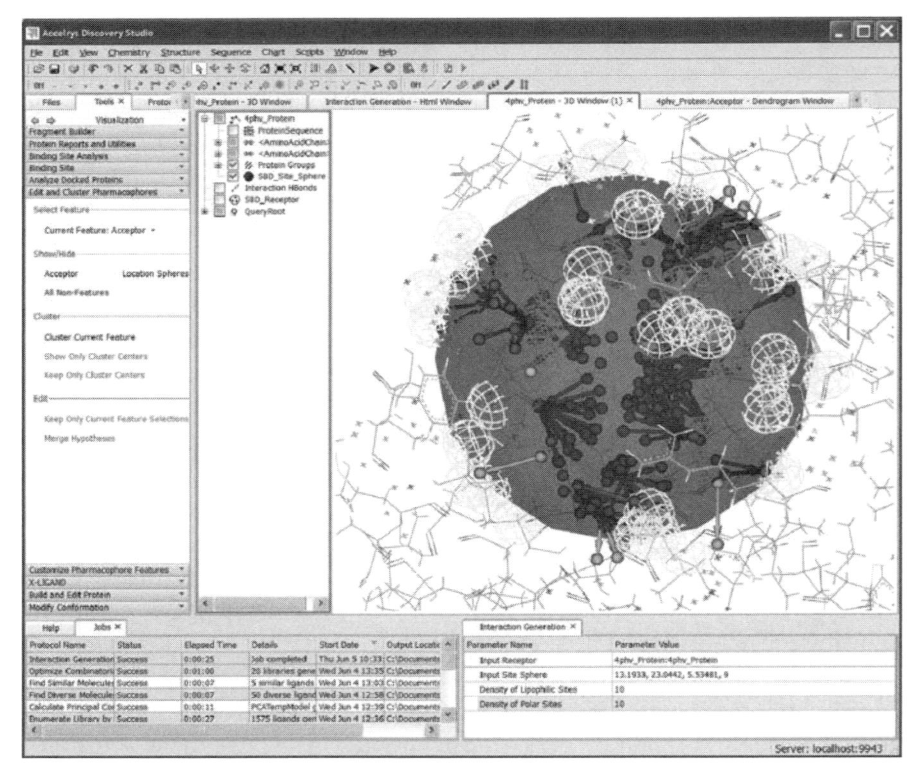


**Figure 13.4** An example of the creation of a pharmacophore from the protein's active site. This example shows the Discovery Studio software from Accelrys, which has a mostly automated process for defining the pharmacophore.

The process of building a pharmacophore model from the geometry of the active site starts with a three-dimensional molecular structure for the protein, or at least the active site. In some cases, it may be necessary to identify which region of the surface of the protein is the active site. There are crevice detection programs to suggest possible active site locations based on the geometry of the surface.

There are automated algorithms to generate a pharmacophore model from the active site. However, it is advisable for a chemist to examine these models and how they interact with the active site residues. It is necessary to look at how far the model extends within the active site and whether it fills specificity pockets, finds the strongest interactions (e.g., hydrogen bonds), etc. It is also possible to design or modify a pharmacophore model manually. The most tedious part of this task can be checking the distances between pharmacophore features and nearby atoms to make sure that the distances are reasonable for those interactions.

Pharmacophore models that describe only the features that an inhibitor should have tend to give many false positives in the search results. This is because they will find compounds that have all of these features, but also a large side group that prevents the compound from fitting in the active site. To avoid this problem, it is recommended that the pharmacophore model include some excluded volume features. These features give a penalty to the molecule's score if atoms are in positions where they would bump into the protein. This maps to the medicinal chemistry concept of a forbidden region.

Nearly all pharmacophore model generation tools assume that there will be no covalent interaction between the compound and the active site residues. If such interactions are desired, it is almost always necessary to build the appropriate pharmacophore model features manually.

If a pharmacophore model is constructed without utilizing activities from known inhibitors, then the model can be used to generate a qualitative scoring metric based on how many features match. Some software packages include the ability to do a curve fit to estimate inhibitor activity from how well the compound fits the pharmacophore. Even at its best, this prediction method tends to be a fairly qualitative estimate.

## 13.4 SEARCHING COMPOUND DATABASES

Once a pharmacophore model has been designed, it can be used to search through a database of compounds. Pharmacophore searching is somewhat slower than using two-dimensional fingerprints, but much faster than doing docking calculations. The result of this will be a qualitative ranking of which compounds fit the pharmacophore best.

The fit between the molecule and the pharmacophore can be used as an alignment device. The pharmacophore searching algorithm translates and rotates each molecule in the Cartesian coordinate system to place it in the best possible alignment with the pharmacophore. It is recommended that pharmacophore search algorithms that check multiple conformers of the compounds be utilized. This gives a set of aligned molecules. This pre-positions the molecules to be automatically run through a docking analysis or a 3D-QSAR program.

The search capabilities and model building capabilities are somewhat different from one software program to another. At the extreme end of this variation is ROCS from OpenEye Scientific Software, which is primarily an analysis of molecular shape. Another unique program is LIQUID, which allows the definition of fuzzy pharmacophore features. Programs such as Catalyst from Accelrys, MOE from Chemical Computing Group, Phase from Schrödinger, DISCOtech from Tripos, and LigandScout from Inte:Ligand are more conventional.

## 13.5 RELIABILITY OF RESULTS

Pharmacophore models are good for finding compounds that fit a number of constraints. The analysis is very fast, making these models ideal for searching very large databases of chemical structures. This tends to be a rather qualitative analysis, because of errors associated with molecule conformation and pharmacophore definition. Docking provides a better vehicle for quantitative prediction of compound activity, at the expense of requiring significantly more CPU time per compound.

One drawback of pharmacophore searching is that it searches databases of three-dimensional chemical structures. 3D structures are those with $(x, y, z)$ Cartesian coordinates for each atom. One of the major sources of error is that some pharmacophore searches analyze the ground state conformation of the potential ligands, or the geometry in the database that is hopefully close to the lowest energy conformation. Thus, there is no guarantee that the biologically active conformation has been checked against the pharmacophore. Searches that analyze a single conformer per molecule are possible, and are sometimes chosen as the best option for searching a very large compound database in a reasonable amount of time. However, analyzing a set of reasonable conformers for each compound usually gives better results, as analyzing only one conformer can fail to find compounds that fit the pharmacophore in a different, but still reasonable, conformation.

Sometimes, a database of 3D conformers is constructed by generating three conformers for each single bond in the molecule. This can result in having a very large 3D database to search, as each molecule is represented by

$3^N$ conformers, where $N$ is the number of nonring single bonds. Alternatively, one conformer could be stored in the database, and others generated as each compound is searched. This technique gives a good compromise between a thorough conformational analysis and one fast enough to be practical for a pharmacophore search. Sometimes, even this is too much to search in a reasonable amount of time, in which case only the lowest energy conformer is used.

This aspect of pharmacophore searches is unlike substructure searches and similarity searches, which can search on 2D chemical structures. 2D structures are those that represent atoms and bonds, such as the pictures created in popular chemical drawing programs. Similarity searches avoid conformation problems by working on the 2D chemical structure. Docking avoids initial conformation problems by performing a rigorous search of possible conformations to find the best one. Pharmacophore searches fall in between, being adversely affected by incorrect conformations, but not having a mechanism to carry out a conformation search.

Another limitation of pharmacophore searches is the potential for error in the way in which the pharmacophore model is constructed, as described in the previous section. This can be alleviated somewhat by keeping hits with one or two deviations from the pharmacophore.

In spite of these limitations, pharmacophore searches have a valuable role in the drug design process. When used in an appropriate manner, they are a useful tool for the drug design chemist to find compounds to be assayed that have a reasonable chance of being active.

## BIBLIOGRAPHY

Alvarez J, Shoichet B. Virtual Screening in Drug Discovery. Boca Raton: CRC; 2005.

Giannakakou P, Gussio R, Nogales E, Downing KH, Zaharevitz D, Bollbuck B, Poy G, Sackett D, Nicolaou KC, Fojo T. A common pharmacophore for epothilone and taxanes: molecular basis for drug resistance conferred by tubulin mutations in human cancer cells. Proc Natl Acad Sci USA 2000; 97: 2904–2909.

Guner OF, editor. Pharmacophore Perception, Development and Use in Drug Design. La Jolla: International University Line; 2000.

Horvath D, Mao B, Bozalbes R, Barbosa F, Rogalski S. Strengths and limitations of pharmacophore-based virtual screening. In: Oprea TI, editor. Cheminformatics in Drug Discovery. Weinheim: Wiley-VCH; 2005. p. 117–140.

Langer T, Hoffmann RD, Mannhold R, Kubinyi H, Folkers G, editors. Pharmacophores and Pharmacophore Searches. Weinheim: Wiley-VCH; 2006.

Milne GWA. Pharmacophore and drug discovery. In: von Ragué Schleyer P et al., editors. Encyclopedia of Computational Chemistry. New York: Wiley; 1998. p. 2046–2056.

Additional references are contained on the accompanying CD.

# 14

# QSAR

The reader has probably already noted that this book contains two chapters on QSAR. This chapter covers conventional QSAR, also called 2D QSAR, 1D QSAR, or just QSAR. The next chapter covers 3D-QSAR. Although the two techniques are conceptually similar, they differ in the process followed and the type of chemical properties that they are used to predict.

QSAR stands for "quantitative structure–activity relationship." The exact same technique is sometimes seen in physical chemistry, where it is labeled QSPR for "quantitative structure–property relationship."

## 14.1 CONVENTIONAL QSAR VERSUS 3D-QSAR

QSAR is conceptually a way of finding a simple equation that can be used to predict some property from the molecular structure of a compound. This is done by using curve fitting software to find the equation coefficients, which are weights of known molecular properties. The molecular properties in a QSAR equation are called descriptors. A descriptor can be any number that describes the molecule. Descriptors can be as simple as the molecular weight, or as obtuse as topological indices. QSAR uses descriptors that are a single number describing some aspect of the molecule. 3D-QSAR uses a

*Computational Drug Design*. By David C. Young

3D grid of points around the molecule, each point having properties associated with it, such as electron density or electrostatic potential.

In general, conventional QSAR is best used for computing properties that are a function of nonspecific interactions between the molecule and its surroundings. For these properties, small changes in molecular structure generally give small changes in the property. For example, conventional QSAR is the method of choice for computing normal boiling points, passive intestinal adsorption, blood–brain barrier permeability, colligative properties, etc. Conversely, 3D-QSAR is better for computing very specific interactions, such how tightly a compound will bind to the active site in one, specific protein. The rest of this chapter discusses QSAR, but not 3D-QSAR.

## 14.2 THE QSAR PROCESS

The QSAR model is an equation for predicting some property from molecular descriptors and coefficients of those variables. The process of creating a QSAR model is similar, regardless of what type of property is being predicted. For this discussion, the value being predicted will be referred to as an "activity."

The first step in creating a QSAR model is to generate a training set of compounds with their experimental activities. Ideally, each of these activities should span the range of possible values for that activity. However, QSAR is better than other techniques (e.g., neural networks) at predicting values that extrapolate beyond the values used to create the model. In order to avoid having the model be skewed by being over-fitted to an unusual compound, it is best to have at least 10 training set compounds for every parameter fit in the QSAR equation. If a sufficient number of experimental activities are known, a small percentage of them may be selected at random to be excluded from the test set, to be used as a validation set.

The next step is to compute descriptors of the test set compounds. QSAR programs often can generate hundreds of different descriptors quickly. Most often, QSAR models are limited to descriptors that can be computed very quickly, such as connectivity indices. This allows the QSAR model to be used to analyze very large databases of compounds. Only in rare case will computationally intensive descriptors be used, such as those obtained from quantum mechanical calculations.

Once descriptors have been computed, it is necessary to select which should be included in the QSAR model. Since most QSAR equations are linear, a correlation coefficient gives a quantitative measure of how well each descriptor describes the activity. Thus, the descriptor with the highest correlation coefficient can be selected. The next descriptor to be selected should be one that correlates well with the activity, but is not strongly correlated to the first descriptor selected (as described by a cross correlation coefficient). This makes it possible to compensate for deficiencies in the first descriptor without redundancy. These metrics provide a basis for selecting descriptors, and often suggest several alternative choices. When manually selecting descriptors, a very useful tool is a correlation matrix, which shows the correlation between each descriptor and the activity and all other descriptors; see Fig. 14.1.

Once the descriptors have been selected, it is a trivial task to generate coefficients of best fit, as this is done internally with a matrix least squares method. The predicted value of the activity for the training set compounds can then be computed and compared with the experimental activities. Analysis of which compounds are not predicted well can sometimes suggest additional descriptors to add to the QSAR equation in order to improve the prediction.

Finally, the activity for the validation set of compounds can be predicted and compared with the experimental activity. This result is the best available indication of how accurately the model can be expected to predict the activity.

When selecting descriptors to include in the model, there are often several seemingly equivalent choices of descriptors available at each step. Therefore, most researchers will typically follow an iterative process of generating several different models to find the best set of descriptors to include.

Correlation Matrix

Database: c:/documents and settings/asndcy01/my documents/sample_pdbs/mao_a.mdb

| | 1 | 2 | 3 | 4 | 5 | 6 | 7 | 8 | 9 | 10 | 11 | 12 | 13 | 14 |
|---|---|---|---|---|---|---|---|---|---|---|---|---|---|---|
| 1. Enzymologic: Ki nM 1 | 100 | -17 | -14 | 33 | -57 | 64 | 33 | -22 | -42 | -12 | 12 | -5 | -2 | -11 |
| 2. a_nH | -17 | 100 | 64 | -15 | 5 | -10 | -5 | -14 | -17 | -25 | -41 | 49 | 40 | 41 |
| 3. b_1rotN | -14 | 64 | 100 | 3 | -7 | 9 | 19 | -13 | -25 | -29 | -29 | 46 | 33 | 25 |
| 4. reactive | 33 | -15 | 3 | 100 | -41 | 54 | 32 | -24 | -17 | 6 | 18 | 12 | 16 | 19 |
| 5. a_nN | -57 | 5 | -7 | -41 | 100 | -50 | 3 | 11 | 84 | 69 | 45 | 43 | 45 | 32 |
| 6. a_nO | 64 | -10 | 9 | 64 | -50 | 100 | 79 | -4 | -31 | 9 | 22 | 5 | 5 | -19 |
| 7. PEOE_PC+ | 33 | -5 | 19 | 32 | 3 | 79 | 100 | -17 | 4 | 38 | 48 | 35 | 29 | -9 |
| 8. lip_don | -22 | -14 | -13 | -24 | 11 | -4 | -17 | 100 | 31 | 19 | -3 | -43 | -27 | -25 |
| 9. vsa_don | -42 | -17 | -25 | -17 | 84 | -31 | 4 | 31 | 100 | 87 | 66 | 32 | 46 | 37 |
| 10. TPSA | -12 | -25 | -29 | 6 | 69 | 9 | 38 | 19 | 87 | 100 | 85 | 44 | 59 | 39 |
| 11. density | 12 | -41 | -25 | 18 | 45 | 22 | 48 | -3 | 66 | 85 | 100 | 42 | 53 | 36 |
| 12. vdw_area | -5 | 49 | 46 | 12 | 43 | 5 | 36 | -43 | 32 | 44 | 42 | 100 | 96 | 81 |
| 13. vdw_vol | -2 | 40 | 33 | 16 | 46 | 5 | 29 | -27 | 46 | 59 | 53 | 96 | 100 | 88 |
| 14. logP(o/w) | -11 | 41 | 25 | 19 | 32 | -19 | -9 | -25 | 37 | 39 | 36 | 81 | 88 | 100 |

Close

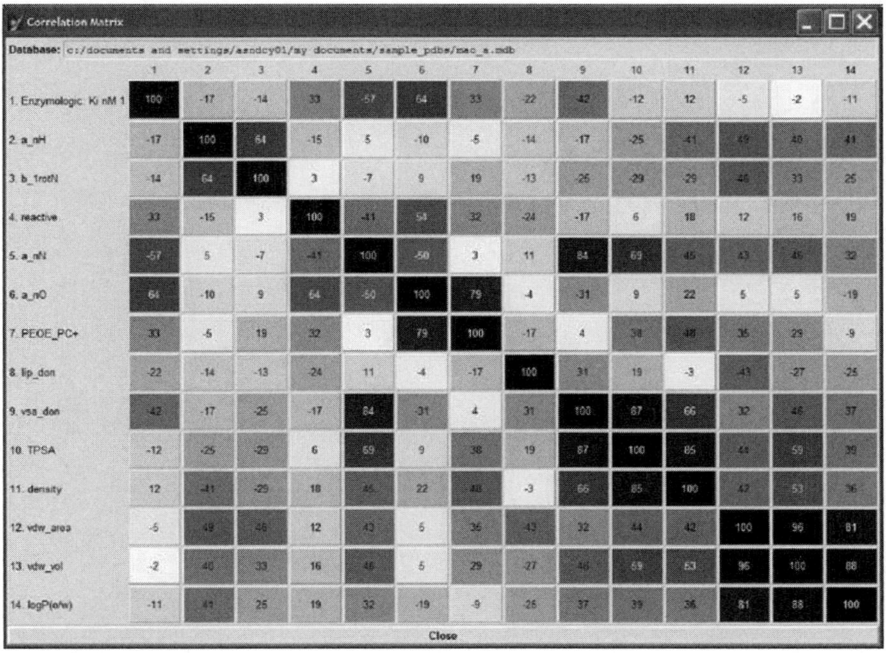

**Figure 14.1** A correlation matrix generated by the MOE software.

Using linear prediction equations and selecting descriptors based on correlation coefficients is a very convenient way to create a QSAR model, and the most frequently used technique. However, it is possible to generate nonlinear equations that contain exponents of best fit, logarithms of descriptors, etc. If there is a known relationship, such as correlation with the square root of a property, that value can be generated and analyzed as just another property in its own right. If the relationship is not known, the researcher can examine graphs of descriptors versus activity in order to visually check which may have a nonlinear relationship. However, this is a more trial-and-error process, as it must be done without correlation coefficients to help guide descriptor selection.

Once the QSAR model has been created, its accuracy can be checked through statistical measures such as the correlation coefficient between actual and predicted values. Often, a few compounds with known results are left out of the training set to be used as a test of the predictive ability. Another surprisingly valuable option is plotting the actual and predicted values, as shown in Fig. 14.2. This plot can be used to identify a nonlinear

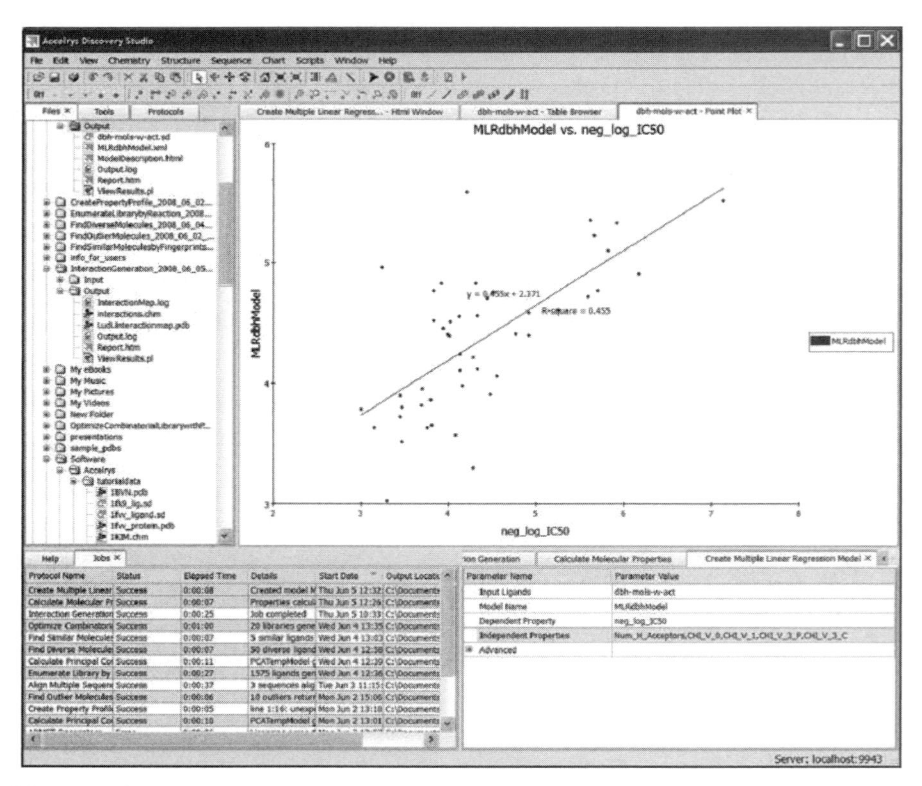


**Figure 14.2** A plot of actual versus predicted results generated in the Discovery Studio software from Accelrys.

trend if it displays a curve. Also, compounds that are outliers can be examined to give insight into the weaknesses of the model, and possibly suggest parameters that should be included to make an improved model.

QSAR software packages differ in the descriptors available, integration with other tools, and ability to fit nonlinear equations. Some are built as several different packages, for example having one interface for the QSAR experts to build QSAR models, and a different interface for laboratory chemists to apply those models to the compounds currently under study.

## 14.3 DESCRIPTORS

Most QSAR modeling is done using descriptors that can be computed very quickly. This allows the resulting equation to be used for very rapid computation of properties for thousands of molecules. However, in some cases, descriptors can be obtained from more time-consuming quantum mechanical calculations. The various descriptors in use can be broadly categorized as being constitutional, topological, electrostatic, geometrical, or quantum chemical.

Constitutional descriptors give a simple description of what is in the molecule. For example, the molecular weight would be considered a constitutional descriptor. Other constitutional descriptors might be the number of heteroatoms, the number of rings, the number of double bonds, etc. Constitutional descriptors often appear in a QSAR equation when the property being predicted varies with the size of the molecule.

Topological descriptors are numbers that give information about the bonding arrangement in a molecule. Some examples are the Weiner index, Randic indices, Kier and Hall indices, the connectivity index, and information content. These are somewhat obtuse in that their values do not have an obvious connection to molecular structure. However, these descriptors do sometimes give the best QSAR predictions, as they can quantify some less concrete concepts, such as whether the molecule is floppy or rigid.

Electrostatic descriptors are single values that give information about the molecular charge distribution. Some examples are polarity indices, multipoles, and polarizability. One of the most frequently used electrostatic descriptors is the topological polar surface area (TPSA), which gives an indication of the fraction of the molecular surface composed of polar groups versus nonpolar groups. Another heavily used descriptor is the octanol–water partition coefficient, which is designated by a specific prediction scheme such as ClogP or MlogP.

Geometrical descriptors are single values that describe the molecule's size and shape. Some examples are moments of inertia, molecular volume, molecular surface area, and other parameters that describe length, height, and width. Note that some of these only provide a reliable molecular description if they are computed with all molecules oriented within the coordinate system in a uniform fashion. These descriptors are often important to the QSAR equation if the property being predicted is dependent upon whether the molecule is more globular or linear.

Quantum chemical descriptors give information about the electronic structure of the molecule. These include HOMO and LUMO energies, refractivity, total energy, ionization potential, electron affinity, and energy of protonation. The HOMO–LUMO gap or ionization potential can be important descriptors for predicting how molecules will react.

Some authors distinguish between regular descriptors, sometimes called 1D descriptors, and 2.5D descriptors. In this terminology, 2.5D descriptors do not give an exact description of the 3D molecular shape, but are dependent upon it. For example, moments of inertia can be considered 2.5D descriptors. Some topological descriptors will give different values for each enantiomer, and may thus be considered 2.5D descriptors. Note that different software vendors may use different classifications for what they term 1D, 2D, 2.5D, or 3D descriptors. These descriptors can give some marginal improvement in QSAR model accuracy, but are not a substitute for 3D-QSAR or pharmacophore models.

## 14.4 AUTOMATED QSAR PROGRAMS

There are a number of software packages with algorithms to automatically select the best descriptors from a long list of available descriptors. These are very valuable, as they might give a more accurate QSAR prediction equation after a few minutes of processing than a scientist would obtain after months of manually selecting descriptors. Because of this, automated algorithms should certainly be used. However, there are some pitfalls to avoid.

The downside of using automated algorithms is that they are purely mathematical. Thus, they may select descriptors that have a correlation with activity by pure random chance, without any scientific basis. This can at times lead to having QSAR equations that are over-fitted. Over-fitted prediction equations tend to reproduce the training set values well, but do poorly on other compounds. Over-fitting is particularly a problem if the size of the training set is smaller than recommended.

There are a number of different algorithms for selecting QSAR descriptors. The most commonly used are regression analysis algorithms. These automate the process of using correlation coefficients and cross correlation coefficients to select descriptors. There are also multivariate analysis algorithms and heuristic algorithms.

Some software packages use a genetic algorithm to try many possible sets of descriptors and select the best one. Genetic algorithms can give very good solutions or poor solutions, depending upon how thoroughly all of the features of the algorithm have been implemented. Thus, it is advised that genetic algorithm analysis be compared with that from other algorithms, at least until the researcher has become familiar with the strengths and weaknesses of the specific genetic algorithm program.

Some software packages have a mechanism for building combinatorial libraries and evaluating them with a QSAR model. It is nice to have this flexibility; however, most combinatorial libraries are designed to maximize drug activity, which is better predicted with docking or 3D-QSAR than with conventional QSAR.

One of the most popular QSAR programs is the Codessa program sold by Semichem. It has a spreadsheet interface with a good set of features for manually selecting descriptors. It also has a number of algorithms for automatically selecting descriptors, which often give very accurate QSAR models. There are also full-featured QSAR packages sold by Accelrys, Tripos, Schrödinger, and the Chemical Computing Group.

## 14.5 QSAR VERSUS OTHER FITTING METHODS

There are a number of techniques that give information that is somewhat similar to that obtained from QSAR calculations. The following paragraphs give a brief discussion of some of the other fitting techniques used in the industry and how they compare with conventional QSAR.

QSAR programs may list their internal algorithm as either a matrix least squares method or as using a linear regression method. These are mathematically equivalent ways of arriving at the exact same answer. Likewise, support vector machines (see Section 17.3) have attracted attention as a statistical learning algorithm for property prediction, but studies have been published indicating that the support vector machine algorithm gives the exact same answer as that obtained from a matrix least squares method. In most cases, the matrix least squares method will execute most quickly on a computer.

Group additivity methods are similar to QSAR in that they allow rapid computation of a molecular property from the chemical structure. These methods break the molecule into a backbone, rings, and functional groups. Each group makes a contribution to the final answer, so the values from all the groups are simply added together. Group additivity methods have been very successful for predicting toxicity, where a single functional group might dominate the toxicity of the molecule. Some of the descriptors commonly used in QSAR equations are computed using group additivity methods. Usually, QSAR equations give more accurate predictions for pharmacokinetic properties.

Neural networks (see Section 17.3) are similar to QSAR equations in that a training set is used to train the network, and then the network is used to predict the property that it was trained to predict. Neural networks can sometimes give more accurate property predictions than QSAR when used as an interpolation method, meaning that the properties of molecules being predicted fall in the range of the properties of the molecules in the training set. Neural networks tend to do much worse than QSAR at extrapolating properties beyond the values of those in the training set.

Hologram QSAR (HQSAR) is a technique in which a molecule's fragments are summed into bins in an array. This array of numerical values acts as a fingerprint, although it is not binary as is a traditional fingerprint. This extended 2D fingerprint is called a "hologram," although it has nothing to do with holographic displays. HQSAR is thus an alternative way of using molecular fragments to predict molecular properties. The HQSAR software is sold by Tripos Inc.

Some decision support programs use the term "QSAR," even though they are not using the mathematical algorithms typically referred to as QSAR. For example, a decision support program might have an option to look at functional groups separate from the rest of the molecule, and define statistical correlations between activity and the presence of certain functional groups. Decision support systems typically utilize clustering, selection, cheminformatics, diversity analysis, and filtering. They may at times have some predictive capability, although it may not actually be the type of QSAR equation described in this chapter.

QSAR is a valuable tool for predicting molecular properties that cannot be computed any other way. In general, QSAR is seldom used for predicting drug activity, as other methods are more accurate for activity prediction. However, QSAR is often the method of choice for predicting pharmacokinetic properties, such as blood–brain barrier permeability and passive intestinal absorption.

## BIBLIOGRAPHY

### Review Articles

Hansch C, Hoekman D, Gao H. Comparative QSAR: Toward a deeper understanding of chemicobiological interactions. Chem Rev 1996; 96: 1045–1075.

Ivanciuc O. Applications of support vector machines in chemistry. Rev Comput Chem 2007; 23: 291–400.

Karelson M, Lobanov VS, Katrizky AR. Quantum-chemical descriptors in QSAR/QSPR studies. Chem Rev 1996; 96: 1027–1043.

Katritzky AR, Lobanov VS, Karelson M. QSPR: the correlation and quantitative prediction of chemical and physical properties from structure. Chem Soc Rev 1995; 24: 279–287.

Kubinyi H. Quantitative structure–activity relationships in drug design. In: von Ragué Schleyer P et al., editors. Encyclopedia of Computational Chemistry. New York: Wiley; 1998. p. 2309–2320.

Peterson K. Artificial neural networks and their use in chemistry. Rev Comput Chem 2000; 16: 53–140.

There are QSAR related resources on the website of the Cheminformatics and QSAR Society at http://www.qsar.org/.

## Books

Karelson M. Molecular Descriptors in QSAR/QSPR (including CD-ROM). New York: Wiley; 2000.

Todeschini R, Consonni V. Handbook of Molecular Descriptors. Weinheim: Wiley-VCH; 2002.

van de Waterbeemd H, editor. Structure–Property Correlations in Drug Research. Austin: RG Landes; 1996.

Additional references are contained on the accompanying CD.

# 15

## 3D-QSAR

3D-QSAR is a technique used for quantitatively predicting the interaction between a molecule and the active site of a specific target. The great advantage of 3D-QSAR is that it is not necessary to know what the active site looks like. Thus, it is possible to use this technique when the target is unknown.

3D-QSAR is a mathematical attempt to define the properties of the active site without knowing its structure. This is done by computing the electrostatic and steric interactions that an imaginary probe atom would have if it were placed at various positions on a grid surrounding a known active compound. In some cases, other interactions such as hydrogen bonding will also be included. After doing this for multiple active compounds, a partial least squares algorithm can be used to determine what spatial arrangement of features there could be in an active site that interacts with the known active molecules.

In the early stages of drug design, when the geometry of the active site is not known, 3D-QSAR tends to be used for more accurate predictions of compound activity. In these stages, pharmacophore searches tend to be more useful for quickly searching very large molecular structure databases, while 3D-QSAR tends to give more accurate activity predictions.

3D-QSAR gives the most accurate results when it is trained on a set of very similar molecules, and used to predict the activity of molecules similar to those in the training set. Pharmacophore models are more forgiving than 3D-QSAR

*Computational Drug Design*. By David C. Young

in this respect, and thus are better for scaffold hopping to find other classes of compounds that also inhibit a given target protein. At later stages in the drug design process, when the geometry of the active site is known, docking replaces 3D-QSAR as the preferred prediction technique. One exception to this pattern is that field-based techniques can perform well for metalloenzyme active sites, which are the downfall of many docking programs.

3D-QSAR is used for predicting a specific interaction, such as the binding of a ligand to the active site of a specific protein. This is unlike conventional QSAR, which is better used to predict general interactions such as colligative properties. It is unfortunate that 3D-QSAR and QSAR bear such similar names, as the two are, for all practical purposes, completely different techniques.

## 15.1 THE 3D-QSAR PROCESS

The first step in building a 3D-QSAR model is to select a training set. This is a set of compounds with their activities. The activities should preferably be from biochemical assays, as these are the experiments that 3D-QSAR should, in principle, model most accurately. 3D-QSAR usually does rather poorly when generating models from *in vivo* data. Ideally, there should be 15–20 compounds or more in the training set, preferably not all built on the same backbone. Training sets with fewer than 20 compounds tend to give less accurate results, owing to outliers skewing the model too heavily. There should be as wide of a range of activity as possible within the training set.

The next step is to generate shapes and alignments of the training set compounds. The desired shape is the shape of the bioactive conformer, which could be up to 10 kcal/mol higher in energy than the ground state conformer. Of course, the bioactive conformer is not known at this stage of the research, so three criteria are used to construct conformers that are the most likely candidates for the bioactive conformer:

- All conformations should be at or near the lowest energy conformation, and certainly no more than 10 kcal/mol above the lowest energy conformation.
- All training set compounds should be put in conformations that give them similar shapes.
- All compounds should be put in conformations that give a close alignment of pharmacophoric features.

This conformer selection and alignment can be done manually, or with automated algorithms that are based either on shape alignment or on alignment of pharmacophoric features. It is possible to perform a pharmacophoric fit, either using a pharmacophore-like description or with a field fitting to align the molecules such that their molecular fields (e.g., electrostatic or steric descriptions) align as well as possible. The shape-based alternative is to use molecular shape analysis (MSA), which optimizes the volume of overlap between two molecular positions. Even when an automated algorithm is used, the chemist will often try various manual adjustments to verify that it seems to be the best possible alignment.

The manual procedure for selecting conformers and performing alignment typically starts with all compounds being put in their ground state conformations. Then, the most rigid molecules are aligned—hopefully with a minimal number of manual conformation adjustments necessary to give a better shape or pharmacophoric fit. Once several of the most rigid compounds have been aligned, this provides a template for shape and pharmacophore feature fit. The compounds that have a similar shape and pharmacophore arrangement in their lowest energy conformation can then be aligned to these compounds. The remaining compounds are aligned by making the necessary adjustments to their conformations to give a reasonable fit to shape and pharmacophoric features.

There are several input parameters that must be set. The grid spacing defaults are usually reasonable, although in highly unusual examples altered parameters may be useful. It is also necessary to select which fields to use. The original CoMFA algorithm used a steric field (Lennard–Jones potential) and an electrostatic field. Some other options are available, including a molecular lipophilic potential (MLP) and a hydrogen bonding potential. There have also been experiments with the desolvation field, that is, the difference between the field in solvent and the field in vacuum. However, the desolvation results have not been shown to differ much from the results with the CoMFA default fields. After performing the ligand alignment, the researcher can examine the dominant features to determine which fields might be most crucial for the particular target. It is advisable to always include the steric field for shape fitting.

The model is created from a large curve fit to find a fitting between fields and biological activity. This is an automated process that is usually done with a partial least squares (PLS) algorithm. The PLS calculation generates a number of PLS components. An analysis must be performed to determine which components should be used to obtain a good fit, without over-fitting. This is almost always done with a leave-one-out (LOO) technique. The software packages provide a good interface for doing this, and good documentation on how to select the best components. The final activity prediction is a linear combination of coefficients times the selected PLS coefficients, with a constant term added in.

## 15.2 3D-QSAR SOFTWARE PACKAGES

The most established 3D-QSAR program is the CoMFA (Comparative Molecular Field Analysis) software package from Tripos, which was created about 20 years ago. CoMFA continues to be the most heavily used 3D-QSAR package. The above discussion is based on CoMFA. The HASL (Hypothetical Active Site Lattice) program is very similar to CoMFA.

The MFA package from Accelrys is very similar to the CoMFA package. The primary difference is that MFA does not sample points on a rectilinear grid. The usefulness of this is the ability to test several different grids in order to avoid incorrect results due to artifacts of the grid point placement. Such artifacts can occur, but do not arise particularly frequently.

CoMSIA (Comparative Molecular Shape Indices Analysis) is a more qualitative, CoMFA-like analysis, also from Tripos. It uses a Gaussian function in place of the typical CoMFA fields. CoMSIA tends to locate features closer to the molecule than CoMFA does. The advantage of this is that CoMSIA can be more discriminating when CoMFA predicts a very large number of compounds to be active. The disadvantage is that CoMSIA may erroneously give a poor property prediction for some compounds that have good activity.

## 15.3 SUMMARY

One of the drawbacks of any QSAR technique is that a reasonable amount of experimental activity data must be available in order to construct the QSAR model. Thus, it cannot be used at the very outset of a project.

Another issue is that the model can only predict based on the space explored by the active compound set. What does it mean if none of the active compounds has a functional group in a given position? It could mean that any candidate with a functional group in that position would be inactive, owing to steric hindrance with one of the active site residues. However, it could also mean that no analogs with functional groups in that position were synthesized and tested. In spite of these limitations, 3D-QSAR is often the most accurate activity prediction method for ligand-based drug design processes.

## BIBLIOGRAPHY

### Books

Kubinyi H, editor. 3D-QSAR in Drug Design: Theory, Methods and Applications. Leiden: Escom; 1993.

Kubinyi H, editor. 3D QSAR in Drug Design: Volume 1: Theory Methods and Applications (Three-Dimensional Quantitative Structure Activity Relationships). Heidelberg: Springer; 2000.

Kubinyi H, Folkers G, Martin YC, editors. 3D QSAR in Drug Design: Ligand–Protein Interactions and Molecular Similarity (Perspectives in Drug Discovery and Design, Volumes 9–11). Dordrecht: Kluwer; 1998.

Kubinyi H, Folkers G, Martin YC, editors. 3D QSAR in Drug Design: Recent Advances (Perspectives in Drug Discovery and Design, Volumes 12–14). Dordrecht: Kluwer; 1998.

## Review Articles

Evans DA, Doman TN, Thorner DA, Bodkin MJ. 3D QSAR methods: phase and catalyst compared. J Chem Inf Model 2007; 47: 1248–1257.

Greco G, Novellino E, Martin YC. Approaches to three-dimensional quantitative structure–activity relationships. Rev Comput Chem 1997; 11: 183–240.

Kubinyi H. Comparative molecular field analysis (CoMFA). In: von Ragué Schleyer P et al., editors. Encyclopedia of Computational Chemistry. New York: Wiley; 1998. p. 448.

Oprea TI, Waller CL. Theoretical and practical aspects of three-dimensional quantitative structure–activity relationships. Rev Comput Chem 1997; 11: 127–182.

Additional references are contained on the accompanying CD.

# 16

# QUANTUM MECHANICS IN DRUG DESIGN

The dominant molecular modeling technique in drug design is molecular mechanics, which is the theory underlying conventional docking algorithms and many other computational drug design tools. Molecular mechanics is used because it is fairly accurate for organic molecules, can model large proteins on typical desktop computers, and can run calculations on a large number of small molecules in a single day. However, molecular mechanics has its limits. There are a number of situations when quantum mechanics is superior to molecular mechanics:

- modeling systems with metal atoms
- increased accuracy
- computing reaction paths
- modeling charge transfer
- predicting spectra
- modeling covalently bound inhibitors
- computing enthalpies of covalent bond formation or breaking

Because of these requirements, quantum mechanics plays a small, but growing role in computational drug design. For example, Fig. 16.1 shows the Trident program, which is designed to be a quantum mechanical program for use in

*Computational Drug Design*. By David C. Young

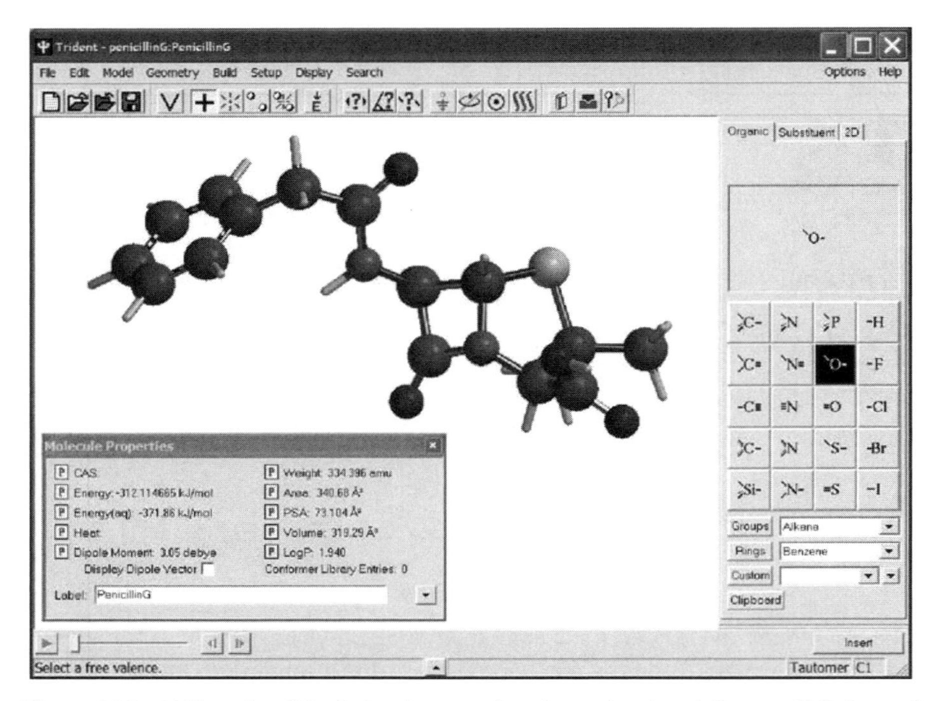


**Figure 16.1** Trident (medicinal chemistry version shown here) and Spartan (full-featured version) from Wavefunction Inc. are programs that can be used to conveniently run quantum mechanical calculations on a desktop computer.

the medicinal chemistry field. Each of the above uses of quantum mechanics is discussed in more detail in later sections of this chapter. The next section presents a discussion of some of the software tools and techniques for accomplishing that task.

## 16.1 QUANTUM MECHANICS ALGORITHMS AND SOFTWARE

Quantum mechanics is the correct mathematical description of the behavior of electrons in atoms and molecules. As such, it is the rigorously correct mathematical method for computing molecular geometries, spectra, enthalpies, etc. The core of quantum mechanics as applied to chemistry is the Schrödinger equation, which is a wave equation involving the kinetic energy of the electrons and Coulombic interactions between electrons and nuclei. The kinetic energy of the nuclei can be treated separately without significant loss of accuracy (this is the Born–Oppenheimer approximation). The existence of quantized electron energy levels (orbitals), chemical bonds, and discrete ionization potentials arise naturally in this approach.

In nonrelativistic quantum mechanics, the ability of each orbital to hold two electrons with opposite spins is put in as a known constraint on the calculation. Relativistic quantum mechanics is required to show the necessity for two electrons of opposite spin. Relativistic quantum mechanics also gives some shifts in orbital size for heavy elements (lanthanides and heavier) due to the increase in mass of electrons as they approach the speed of light near a highly charged nucleus. This effect can also be included in a nonrelativistic treatment through the use of effective core potentials to model the inner shells.

Unfortunately, to date, the Schrödinger equation has only been solved exactly for systems with one electron. In the absence of exact solutions for multielectron systems, it is necessary to use approximation schemes. Many such schemes are available, including Hartree–Fock theory (HF), density functional theory (DFT), Møller–Plesset perturbation theory (MP2, MP3, etc.), coupled cluster (CC) theory, configuration interaction (CI), semiempirical schemes, and many more. Having so many approximation schemes makes it possible to perform a quantum mechanical calculation to any accuracy desired. Unfortunately, there are diminishing returns in that performing higher accuracy calculations requires exponentially increasing amounts of computer CPU time, memory, and disk space.

Semiempirical methods run much faster than *ab initio* quantum mechanical calculations. Calculations that are based solely on the quantum mechanical equations and some approximation method with which to solve those equations are referred to as *ab initio* calculations. Semiempirical calculations keep the overall structure of quantum mechanical methods, but replace the most time-consuming parts of those calculations with values parameterized from experimental results. This makes semiempirical calculations much faster to run, at the expense of some loss of accuracy. Semiempirical calculations can be very valuable tools, as long as the researcher knows how to utilize them appropriately.

There have been cases where quantum mechanical calculations have given incorrect results owing to the selection of one of the more rudimentary approximation schemes. However, when sufficiently accurate approximations are used to ensure that the correct solution to the Schrödinger equation is being obtained, quantum mechanics has never been wrong. Like Newton's laws of motion, quantum mechanics is a law of nature, and thus a perfect mathematical description within its scope of applicability. Indeed, the correspondence principle is illustrated as a classroom exercise in which students show that substituting large values for the masses in the quantum mechanical equations results in certain terms becoming so immeasurably small that all that remains are Newton's equations.

The role of quantum mechanical methods in drug design is increasing, although they are still underused. The primary restriction that is preventing quantum mechanics from being adopted more quickly and more widely is the fact that quantum mechanics is much more computationally intensive than molecular mechanics. Even for a modest number of atoms, a quantum mechanical calculation could take hours where a molecular mechanical calculation takes seconds. The fact that many industrial drug design departments must model thousands of compounds per day makes this a significant issue. Fortunately, there are several technologies that are lowering this barrier to adoption of quantum mechanical methods.

A number of linear scaling algorithms have been created for quantum mechanical methods. This has allowed the algorithmic time complexity to be lowered, at the very best from $O(N^4)$ to $O(N)$ for sufficiently large molecular systems. These algorithms do not really change the time necessary to model small compounds. However, they can be essential if it is necessary to model a significant portion of a protein with an all-electron *ab initio* method. Even with these techniques, a calculation on an entire protein may be prohibitive, thus making it necessary to model only the section of the protein around the active site. Linear scaling algorithms have been included in a number of quantum mechanical software packages, such as the Gaussian software, Q-Chem, and DivCon from QuantumBio, Inc.

Algorithmic improvements have been necessary to get even semiempirical techniques for modeling entire proteins. The MOZYME derivative of the MOPAC program uses localized orbitals to make semiempirical calculations on entire proteins feasible, sometimes even on desktop computers.

Probably the most widely used techniques for applying quantum mechanics to drug design are QM/MM methods, which combine quantum mechanics and molecular mechanics in the same calculation. The key part of the system, such as where bonds are being formed and broken in enzyme catalysis, is modeled using a quantum mechanical method, while atoms beyond some range are modeled by molecular mechanics. It takes some understanding of the computational method and the chemistry of the system to select a reasonable boundary between the methods. There are several QM/MM algorithms in use, differing in the way in which they interconnect the two computational models, and they may allow three concentric regions modeled by quantum mechanics, a semiempirical method, and molecular mechanics. There are QM/MM methods in Gaussian, GAMESS, and the QSite module from Schrödinger, Inc.

When all else fails, quantum mechanical calculations can be used to parameterize a molecular mechanics force field for one specific task. Often, this is done manually, by someone who is familiar with the mathematics of how force fields are parameterized. Note that parameters cannot be taken out of one force field and put into another. There are a few automated algorithms for this type of parameterization. For example, the QPLD module from Schrödinger performs QM-polarized docking. This means that a quantum mechanical calculation is used to determine the partial charges on the atoms, which are in turn used as part of a molecular mechanics calculation. This both improves the accuracy of the electrostatic portion of the molecular mechanics calculation and allows the full simulation to run with the speed of a molecular mechanics calculation.

## 16.2 MODELING SYSTEMS WITH METAL ATOMS

Molecular mechanics methods work well for organic molecules primarily because the atoms in such molecules have well-defined hybridization states that exhibit little change in preferred bond angles or charge transfer when other species interact noncovalently (i.e., via hydrogen bonding or van der Waals interactions). Metal atoms, however, change their coordination sphere and undergo significant electron transfer when another species comes near them. This is, of course, why metals make such good catalysts, but it is also why molecular mechanics generally describes systems with metal atoms far less accurately than completely organic systems.

Some molecular mechanics force fields model metal atoms as spheres with a charge. Others parameterize metal atoms for one particular oxidation state and coordination. Both of these models are fundamentally poor descriptions, as metal atoms frequently shift coordination and transfer charge as a ligand enters the active site of a metalloenzyme. Quantum mechanics, by its very nature, can do a much better, more accurate, job of modeling a metal atom's fluidity of hybridization and charge. Thus, the potential applicability of quantum mechanics to drug design is at least as great as the number of drug targets with metal atoms in the active site.

Quantum mechanics is definitely the best option for modeling metal atoms from the standpoint of obtaining results that are qualitatively and quantitatively correct. Unfortunately, metal calculations—particularly for transition metals—are particularly difficult tasks from a technical standpoint. When modeling transition metal compounds quantum mechanically, the researcher should expect to encounter many technical difficulties. These include failure to converge, converging to the wrong electronic state, inclusion of relativistic effects, and very long run times (quantum mechanical calculations scale as the number of electrons to some power). In such cases, it is advisable to obtain the services of a researcher who has spent their career doing complex quantum mechanical calculations.

## 16.3 INCREASED ACCURACY

The QPLD example mentioned earlier in this chapter is just one way in which quantum mechanical calculations can be utilized in drug design to give more accurate results. There are a number of ways in which quantum mechanical methods are used to improve accuracy, even in systems that do not contain metal atoms.

Molecular mechanics calculations do not allow chemical bonds to be formed or broken over the course of the simulation. As such, quantum mechanics is often better for modeling suicide inhibitors, which bind irreversibly to the target's active site. There are also drugs that have high potency because they bond covalently to the active site, albeit in a reversible reaction. These scenarios are modeled more accurately with quantum mechanical methods, which give reasonable bond energies, than with molecular mechanics, which will only find a hydrogen bond or van der Waals interaction in a typical docking simulation.

Some biological systems undergo electron transfer reactions. Again, molecular mechanics does not explicitly model the electron charge distribution, and thus cannot model charge transfer without some manual intervention to shift atomic charges. Charge transfer is readily simulated by quantum mechanical methods.

Quantum mechanical methods have also been used to generate QSAR descriptors. Often, these are values unique to quantum mechanical calculations, such as the HOMO and LUMO energies. QM can also be the best option for computing descriptors related to the electron density or electrostatic potential of the molecule.

Quantum mechanical methods can give highly accurate predictions of acidity. This can be particularly important for molecules with unusual structures, for which empirical $pK_a$ prediction is less reliable.

In some drug design projects, quantum mechanical methods are used to obtain very accurate molecular geometries and relative energies of conformers. Molecular mechanics generally gives accurate geometries and conformer energies for organic molecules. However, there are compounds that are not predicted well by molecular mechanics. Even if the molecular mechanics is working well, the maximum possible accuracy may be desired at later stages of the drug design process.

The Carbó Similarity Index (CSI) is a quantitative measure of how similar to molecules are to one another. It is derived from the Quantum Similarity Measure (QSM). This can be used as a metric for aligning molecule, such as in preparation for a 3D-QSAR or pharmacophore study. This can be an improvement over other alignment techniques when the compounds have similar electron density distributions, but do not share a common backbone.

The single most heavily used, and thus important, computational technique in drug design is docking. Conventional docking algorithms can be considerably less accurate than desired, particularly when there is a metal atom in the target's active site. This has led to the development of several quantum mechanical docking and scoring techniques. The only quantum mechanical docking code that is available on the open market is the QUANTUM docking program from Quantum

Pharmaceuticals. Other companies have kept their technologies for in-house use, such as the QM/MM docking code developed at Exegenics and the quantum mechanical free energy perturbation technique developed at Metabasis.

## 16.4 COMPUTING REACTION PATHS

A significant percentage of drug targets are enzymes that catalyze some chemical reaction. In order to create a drug that inhibits an enzyme, it is important to understand the mechanism of the reaction that the enzyme undergoes with the native substrate. Reaction mechanisms can be understood at various levels of detail. The organic chemist's view of pushing electrons around on paper can often give a correct and useful view of the reaction mechanism. Quantum chemical calculations give a more detailed view of how molecules are oriented, and how electron transfer is (usually) simultaneous with bond lengthening or shortening.

In modeling electron transfer reactions, quantum mechanical methods can display charge transfer in several different ways. If a single electron is moving, plots of the spin density can show the unpaired electron density as a 2D or 3D plot. If it is a singlet spin system, density difference or molecular electrostatic potential (MEP) plots can show the electron movement. Total electron density plots are the least useful, because the electron density shift due to bond formation is such a tiny fraction of the total electron density.

Reaction mechanisms computed by quantum mechanical methods are usually intrinsic reaction coordinates (IRC). Such computations reflect the minimum energy path between reactants and products. In reality, the atoms are vibrating, and the reaction can be slightly different as there are many energetically accessible paths, with the IRC being the most energetically favorable. Again, the size of the molecule often makes it necessary to model just the atoms near the active site, or use a QM/MM method.

Quantum mechanics is also the most accurate way of obtaining bond enthalpies by taking the difference in total energy between reactants and products. These calculations can be used by themselves, or as the starting point for obtaining the entire reaction coordinate.

## 16.5 COMPUTING SPECTRA

Quantum mechanical calculations are the most accurate means for computationally predicting spectra, including infrared, Raman, ultraviolet–visible, microwave, and NMR spectra. Mass spectra are better computed with tools designed specifically for that purpose, and even then may not be predicted very accurately.

Infrared spectra are computed using a frequency calculation to obtain the normal modes of vibration. Some software packages also perform the transition dipole moment calculation to give the intensities of the spectral peak. These intensities can show which modes are infrared-active and which are Raman-active. Quantum mechanical calculations work well for obtaining the infrared spectra of small molecules. For large proteins, the intensive nature of these calculations makes them intractable even on the largest computers available at present. A better way to obtain vibrational information about large biomolecules is to carry out a large molecular dynamics simulation followed by a Fourier transform from the time domain to the frequency domain.

The entire ultraviolet–visible spectrum of a molecule can be estimated from a single quantum mechanical calculation at the configuration interaction single excitation (CIS) level. More accurate results can be obtained by performing separate quantum mechanical calculations on each electronic excited state and ionized form of the molecule. There are also variations on the CIS algorithm designed specifically for improved electronic spectra prediction, such as those in the Q-Chem software.

Some quantum mechanics programs include algorithms for computing NMR chemical shifts, as well as other magnetic properties. For example, the Gaussian and GAMESS programs can compute magnetic properties.

Although quantum mechanical techniques cannot be used for entire proteins, they serve a valuable role in providing information that no other method can give. The long-term trend has been for more types of systems to come under the quantum mechanical umbrella as computer hardware and quantum mechanical software allows.

## BIBLIOGRAPHY

### Books

Carbo-Dorca R, Robert D, Amat L, Girones X, Besalu E. Molecular Quantum Similarity in QSAR and Drug Design. Berlin: Springer; 2000.

Young D. Computational Chemistry: A Practical Guide for Applying Techniques to Real World Problems. New York: Wiley; 2001.

### Review Articles

Besalu E, Girones X, Amat L, Carbo-Dorca R. Molecular quantum similarity and the fundamentals of QSAR. Acc Chem Res 2002; 35: 289–295.

Peters MB, Raha K, Merz KM. Quantum mechanics in structure-based drug design. Curr Opin Drug Discov Dev 2006; 9: 370–379.

## Reference Specific to QM/MM

Gao J. Methods and applications of combined quantum mechanical and molecular mechanical potentials. Rev Comput Chem 1996; 7: 119–185.

## References Specific to Intrinsic Reaction Coordinates

Bernardi F, Oliviucci M, Robb MA. Potential energy surface crossings in organic photochemistry. Chem Soc Rev 1996; 25: 321–328.

Dunning TH, Harding LB. Ab initio determination of potential energy surfaces for chemical reactions. In: Baer M, editor. Theory of Chemical Reaction Dynamics. Volume 1. Boca Raton, CRC Press; 1985. p. 1–69.

Kuntz PJ. Semiempirical potential energy surfaces. In: Baer M, editor. Theory of Chemical Reaction Dynamics. Volume 1. Boca Raton, CRC Press; 1985. p. 71–90.

McKee ML, Page M. Computing reaction pathways on molecular potential energy surfaces. Rev Comput Chem 1993; 4: 35–65.

Sidis V. Diabatic potential energy surfaces for charge-transfer processes. In: Baer M, Ng CY, editors. State-Selected and State-to-State Ion-Molecule Reaction Dynamics. Part 2: Theory. New York: Wiley; 1992. p. 73–134.

Simonetta M. Organic reactions paths: A theoretical approach. Chem Soc Rev 1984; 13: 1–14.

Truhlar DG, Steckler R, Gordon MS. Potential energy surfaces for polyatomic reaction dynamics. Chem Rev 1987; 87: 217–236.

Additional references are contained on the accompanying CD.

# 17

# *DE NOVO* AND OTHER AI TECHNIQUES

Artificial intelligence (AI) is the study of computer algorithms for doing things that would typically require human intelligence. There is no exact definition of AI, and over the years some things that were originally designed by AI researchers have become mainstream tools, and subsequently are no longer considered part of the AI field. Some tasks currently considered part of AI are programs that can learn, draw logical conclusions, exhibit some level of understanding, or communicate in written or spoken human languages.

Some of the initial work that led to the development of spellcheckers, grammar checkers, and web search engines was originally part of AI research to understand languages and knowledge, but is not generally considered part of AI today. AI researchers are currently debating whether symbolic manipulation (the technology inside Mathematica and Maple, for example) and neural networks should continue to be considered part of the AI field. When the field of AI was conceived, it was envisioned as the creation of a computer program that behaves as though the operator were talking to a real person, who just happened to be sitting inside the computer (in those days, computers were larger than many apartments!). Some researchers are still working towards that goal. However, as the above examples illustrate, most technologies created by AI researchers have thus far just become absorbed in ubiquitous applications.

Likewise, there are drug design software packages that look like any other software package, but are differentiated technologically by the internal use of

*Computational Drug Design*. By David C. Young

AI algorithms. This chapter discusses a number of these applications, first describing what the software does, and then giving a brief conceptual discussion of how the internal algorithm works.

## 17.1 *DE NOVO* BUILDING OF COMPOUNDS

*De novo* programs attempt to automate the process of structure-based drug design. More specifically, these programs design compounds to fit in a particular active site, or conform to a particular pharmacophore model. This is an admirable goal, although it is also an extremely complex task. In practice, *de novo* programs have become part of the computational drug design toolbox. These programs can create some compounds that bind to the target, and may suggest classes of compounds that the researchers had not considered. However, it is seldom, if ever, that the compounds created by *de novo* programs are the final optimized structures that go on to clinical trials.

The first *de novo* programs were very special purpose ones. For example, some of them could build new compounds—but only peptides. The current generation of *de novo* programs build structures either from individual atoms or from small functional groups, thus allowing a large variety of compounds to be constructed. Although the details of the algorithms vary significantly, all of the current *de novo* programs follow three basic steps:

1. Analysis of the active site
2. Building molecules
3. Sorting/selecting for the best candidates

There are algorithms for finding the active site of the protein, but this is usually done in a separate step, not as part of *de novo* compound generation. The active site analysis step gives the molecule-building program information about regions in the active site where there are hydrogen bond donors and acceptors, and other features. This is sometimes done by creating a pharmacophore-like description of the active site. It may also be done by generating a grid interaction potential, just as is done for grid-based docking programs. Some *de novo* programs just use a molecular mechanics-style model of the active site, and then compute interactions between various functional groups and the active site at every step of the process.

The algorithm at the core of *de novo* software generates new compounds by putting together molecular fragments, such as known functional groups and carbon atoms. Two problems with this process are immediately apparent. First, the number of possible compounds that could be generated from a large fragment list is intractably large, making it impossible for the computer program to simply generate every possible compound and test them all. Second, many structures that could come about from randomly connecting fragments would not be chemically sensible (stable and synthetically accessible). The purpose of AI algorithms in *de novo* programs is to address these issues. Most of these AI algorithms in *de novo* software are heuristics, an intelligent process for making reasonable choices at each step. Some are also genetic algorithms, or closely related evolutionary algorithms.

There are a number of different algorithms for building molecules. One algorithm consists of docking many molecular fragments into the active site, and then linking these fragments together. The LUDI software package from Accelrys uses this type of fragment-joining algorithm. Some software packages, such as the SPROUT program shown in Fig. 17.1, have graphical interfaces designed to guide users through the sequence of steps necessary to follow this process.

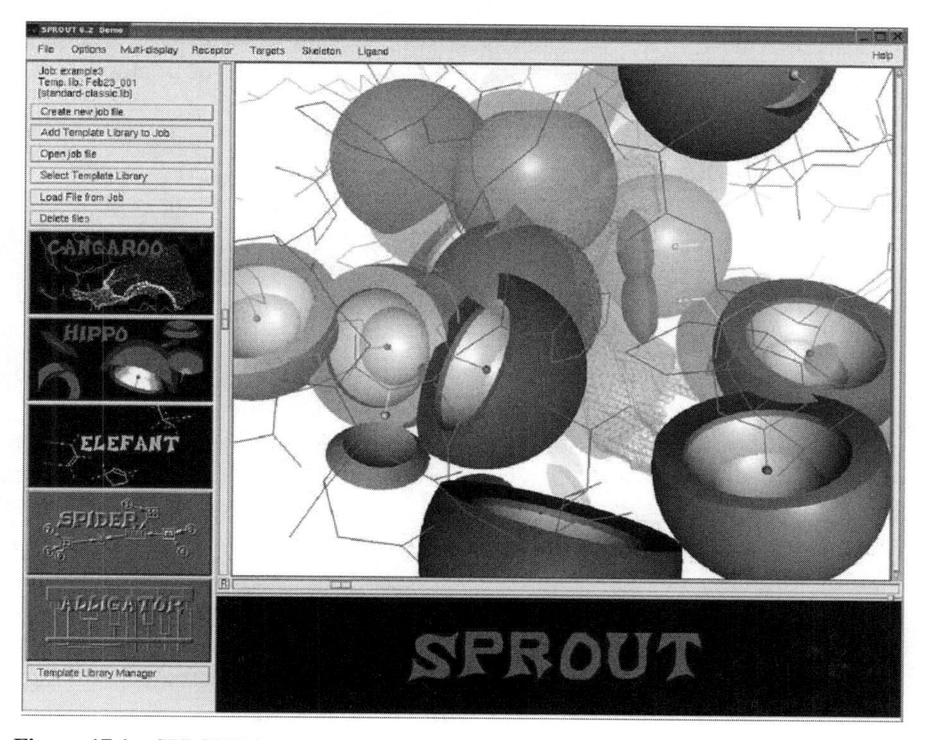


**Figure 17.1** SPROUT is a *de novo* program from SimBioSys. It has a graphical interface designed to lead the user through the processes of setting up and running a *de novo* design process.

Another algorithm in use is a chain growth algorithm in which a molecule is built up one functional group at a time from a starting fragment (synSPROUT from SimBioSys uses this type of algorithm). Once the starting point for a chain growth algorithm is set, atoms or functional groups are added to the structure. As each group is added, an analysis is performed to determine which groups are optimal in that position, and their conformational orientation. Often, a number of functional groups that score well will be selected. In the next step, the program will add another functional group to every one of the trial solutions generated in the previous step. Thus, the number of trial solutions increases with each step of the process. The selection of which intermediate trial solutions to keep may be made on the basis of a docking-type simulation and scoring, or a fit to the active site pharmacophore model, or, in some cases, the steric fit to the active site only. A few *de novo* programs include a secondary selection criterion, such as examining the ADMET properties of the molecule.

Some of the molecule-building algorithms are deterministic, so that they would give the exact same results if the same calculation were run again. Some are stochastic, utilizing a Monte Carlo analysis at some step in the process. Although stochastic methods do not necessarily give the same results every time, the results from multiple runs should be similar. If stochastic methods give significantly different results from one run to the next, it usually means that the number of iterations for stochastic optimization has been set too small.

The chain growth algorithm tends to give chemically sensible compounds, but may have difficulties in regions where the atoms only link, thus giving no logical choice of a fragment to interact with the active site. The linking algorithm tends to find solutions with more rings, and thus more drug-like compounds. Typically, all of the algorithms available will be used to suggest classes of compounds that could be explored.

Most of the molecule-building algorithms are single-pass, meaning that the functional groups are not modified once they have been added to the structure. There are also two-pass algorithms that first build a backbone of carbon atoms, and then add or modify groups to maximize electrostatic interactions with the active site.

In order to achieve the third criterion of selecting the best candidates, the program must be able to address two concerns: first, a way to quantify which candidate structures are best; second, a means for negotiating a combinatorial explosion of possible candidates. One part of the selection process is often designed to remove from consideration compounds that are not suitable. These might include compounds that will not be bioavailable, or compounds that fail a computed synthetic accessibility criterion.

Compounds are sorted on the basis of some estimation of which will make the best drugs. This is sometimes done by performing a conventional docking analysis on every compound. It may also be done by fitting the shape of the active site. Some programs, such as PRO_LIGAND, utilize a pharmacophore description of the active site.

A variation on the *de novo* idea is represented by the SPROUT-LeadOpt program from SimBioSys. Rather than designing a compound to fit in an active site, SPROUT-LeadOpt designs new compounds to be similar to identified active compounds.

There is no consensus as to which *de novo* program is best. For that matter, it is not easy to define what makes one program better or worse than another. Often, multiple programs are used for a given target. Which gives useful results can vary from one target to the next.

## 17.2 NONQUANTITATIVE PREDICTIONS

Nonquantitative predictions, or non-numerical predictions, are results from computer program that are not in the form of numbers. These include results from programs that generate a synthesis route or categorize compounds by their level of toxicity. In some cases, a nonquantitative algorithm might be used to select between a number of available numerical prediction models. Another example is a dialogue-based teaching program in which the student can learn about chemistry by interacting with a computerized dialogue program, just as though they were conversing with a real person via a computer chat system.

There are a number of types of algorithms that may be used inside these programs. The following paragraphs discuss some of these algorithms.

The classic expert system algorithm is a series of decisions. After each decision has been made, the algorithm moves to the next step, which is either another decision or a result. Typically, broad categorization is done by the decisions at the beginning of the process, while much more specific points are addressed later in the decision structure. For example, a program to analyze molecular toxicity might make the first choice based on whether there are any metal atoms in the compound, and near the end of the process it be looking for the presence of groups that have a more subtle effect, such as acetyl groups. Once the decision structure, called a tree, gets to a point at which it cannot go any farther, called a leaf node, the final action is taken. This might be as simple as printing the words “highly toxic” or as complex as running a specific numerical prediction model. The DEREK program uses an expert system to predict whether a compound will be very toxic, slightly toxic, etc. The REX program is used to generate new rules to be added into DEREK’s database.

A similar format is a heuristic algorithm. Rather than having a hard-coded decision structure, a heuristic algorithm will be given rules, or general guidelines for how to proceed. For example, heuristic-based spectral interpretation programs such as heuristic-DENDRAL will be given rules that indicate a general strategy for analysis of a mass spectrum into a structure prediction. DENDRAL has been augmented with learning algorithms that extend the program's capabilities by feeding in spectra–structure pairs. More recently developed spectral analysis programs such as SESAMI, and CHEMICS extend the original DENDRAL method by adding the ability to utilize NMR data in addition to mass spectral data.

Some AI programs function primarily by searching through a database of stored information. For example, the programs OCSS, LHASA, and SECS search a database of chemical reactions. In the language of AI, these reactions transform the system from one state (the reactants) to another state (the products). An algorithm that uses this database lookup format is called a state space search program. These and other synthesis route planning programs have, in more recent versions, been incorporating various types of synthesis strategy as an improvement to augment the more trial-and-error search techniques. Some are now retrosynthetic programs, which work backwards from a desired product to commercially available reagents.

Some software packages use Bayesian inference. This is a statistical process in which many possible hypotheses are set forth, and then the best are selected by utilizing Bayes' theorem. This is a technique for selecting rules for chemical behavior that occur significantly more often than would be expected based on random chance.

Not all decisions are a cut-and-dried yes/no answer. For example, a doctor diagnosing a patient with a particular symptom may know only that the symptom occurs in 80% of patients with one disease and 5% of patients with another disease. Even when all symptoms are taken into account, there may be multiple possible causes. In this case, the doctor must select a course of action based on his understanding of the probability of the various conditions occurring with the observed severity of symptoms. In this example, the doctor is dealing with many "maybe" results, and only marginally quantifying this with probabilities. Fuzzy logic algorithms arrive at a conclusion using this type of probability-based reasoning.

Another difference between programs is the origin of the data used to create the program. There are three sources of data to consider:

- information obtained from questioning an expert in the field
- information based on observation of experimental data
- information obtained through logical reasoning based on the underlying scientific principles

When rules are based on a subject matter expert, the most difficult part of the process is extracting the information from the expert. There is a whole subdiscipline of knowledge engineering devoted to the extraction of knowledge, and putting it into a format usable by the AI program.

Rules based on observation of experimental data are a double-edged sword. These are easier to implement than other algorithms. They may also give the most accurate predictions. However, incorrect results may arise owing to misinterpretation of the data. These are typically artifacts of the data, rather than rigorous principles. For example, a program automatically generating knowledge about families may conclude that the tallest person in the family is the father. Likewise, a program analyzing the personal biographies in an encyclopedia came up with the observation that most people are famous.

Rules based on causal learning are derived from fundamental principles. This is attractive in principle, as this type of reasoning is often applied by scientists. It also has the advantage of not generating wrong answers due to statistical anomalies in the data. The disadvantage is that a useful result may not be obtained if the program does not have a sufficient reasoning capability or sufficient information in its knowledge base. This is probably the most complex type of AI algorithm to implement. Because of the difficulty of implementation, causal learning is the least frequently utilized of these techniques.

## 17.3 QUANTITATIVE PREDICTIONS

There are many ways to make quantitative predictions of chemical properties. For many colligative properties, QSAR (quantitative structure–activity relationship) is the best technique for making numerical predictions. Various types of structure–activity relationships are also used in drug design. Although QSAR is a mathematical method, it often requires a rational analysis by a chemist to find the best combination of descriptors to give an accurate QSAR equation. To help address this problem, an expert system, called APEX-3D, was designed specifically to aid in generating QSAR models.

Another type of quantitative prediction is 2D-to-3D structure conversion. This means converting 2D chemical structures consisting of connectivity and bond order data (e.g., in SMILES or SLN format) to 3D structures with Cartesian coordinates for each atom. Two popular programs for doing this are CONCORD and CORINA. At their core, these programs are rule-based systems. For example, they have rules for selecting the most stable conformation for a cyclohexane ring, unless that conformation is made unfavorable owing to steric hindrance. Some portions of this work are done by using look-up tables of preferred bond distances and angles, which can be executed more rapidly than queries against a large database of group geometries. CORINA will, in certain instances, make adjustments by utilizing a Hückel molecular orbital method. CONCORD and CORINA depend upon collections of hand-curated conformational data, whereas the AIMB program generates its own data collection through analysis of crystallographic coordinate databases.

Work has been done on developing programs that are theoretically capable of deriving new scientific laws. For example, the BACON program re-derived the ideal gas law from raw data. This is an interesting task from the standpoint of considering where AI could potentially go in the future. However, such programs have not contributed significant new scientific laws to the field. More limited learning programs have been more successful, such as those that learn to do conformational analysis or assign spectra from data.

Decision trees are hierarchical categorization algorithms. The decision tree is created with a learning algorithm that selects the best criteria for separating compounds with high values from those with low values at each step of the analysis. Once the tree has been created, unknown compounds can be put into the algorithm, and will be categorized until they are assigned to a leaf node. At the leaf node, the value assigned may be the average of the training compounds at that node, or it may run a specialized prediction model for the compounds at that node, such as a QSAR calculation.

Neural networks are algorithms that make predictions by simulating the way in which brain cells are believed to talk to one another. The network is first trained with a set of experimental data values for a set of compounds, and can then be used to generate predictions for new compounds. Neural networks are powerful interpolation tools, but poor extrapolation tools. That is, neural networks can make accurate predictions of some chemical properties if the unknown molecule's property value falls within the range of those of the training set compounds, but if the value is outside the range of the training set, they generally give very poor predictions. Neural networks have been used for a number of physical chemical properties, such as the normal boiling point.

A related topic is that of support vector machines. A support vector machine is a learning algorithm that gives an end result equivalent to a linear regression or matrix least squares technique. It is classified as an AI technique because it uses a learning algorithm to train the system, which is then used to make the prediction.

Some authors categorize genetic algorithms under artificial intelligence. In this book, genetic algorithms (GA) are viewed as global search techniques. This is primarily because genetic algorithms depend upon natural selection, rather than rational or statistical decision making.

The examples given in this chapter are typical of how algorithmic developments from the AI community have been applied to chemistry and many other fields. Most of these programs do not, on outward appearances, seem to be intelligent. However, they are distinguished as AI applications because their internal mechanisms utilize a knowledge base, learning algorithm, neural network, etc. Although not the core of computational drug design, these techniques still provide enough value to be worth utilizing.

## BIBLIOGRAPHY

### Books

Cartwright HM. Applications of Artificial Intelligence in Chemistry. Oxford: Oxford Science Publications; 1993.

Cartwright HM, Sztandera LM. Soft Computing Approaches in Chemistry (Studies in Fuzziness and Soft Computing). Berlin: Springer; 2003.

Cartwright HM. Using Artificial Intelligence in Chemistry and Biology: A Practical Guide (Chapman & Hall/CRC Research Notes in Mathematics). Boca Raton: CRC Press; 2008.

Hippe Z. Artificial Intelligence in Chemistry: Structure Elucidation and Simulation of Organic Reactions (Comprehensive Chemical Kinetics). Amsterdam: Elsevier; 1991.

Vannucci M, Do KA, Müller P. Bayesian Inference for Gene Expression and Proteomics. Cambridge: Cambridge University Press: 2006.

### Review Articles

Böhm HJ, Fischer S. *De novo* ligand design. In: von Ragué Schleyer P et al., editors. Encyclopedia of Computational Chemistry. New York: Wiley; 1998. p. 657–663.

Clark DE, Murray CW, Li J. Current issues in *de novo* molecular design. Rev Comput Chem 1997; 11: 67–125.

Johnson AP, Green SM. *De novo* design systems. In: von Ragué Schleyer P et al., editors. Encyclopedia of Computational Chemistry. New York: Wiley; 1998. p. 650–657.

Murcko MA. Recent advances in ligand design methods. Rev Comput Chem 1997; 11: 1–66.

Additional references are contained on the accompanying CD.

# 18

# CHEMINFORMATICS

Cheminformatics is the handling of chemical information in electronic form. The term is generally applied to systems for handling large amounts of chemical data. These include databases, chemical structure representation, various ways of searching chemical data, and techniques for the analysis of large collections of chemical data. Cheminformatics is also sometimes called "chemoinformatics" or "chemical informatics." Some texts classify pharmacophore searching, QSAR, and combinatorial library design as part of cheminformatics, because these utilize some cheminformatics algorithms internally. This book has separate chapters devoted to these topics, so they will not be addressed in this chapter.

Cheminformatics is a field in which there is a disparity in the chemistry community. Cheminformatics is very important in the pharmaceutical industry, but seldom taught in the academic chemistry curriculum. In fact, the majority of cheminformatics tools are seldom used in academic research settings. This is because academic research projects typically focus on a small group of compounds, perhaps dozens, whereas researchers in the pharmaceutical industry may have to negotiate databases of millions of compounds. With the current generation of combinatorial chemistry and high throughput screening techniques, it is not unusual for a single drug design project to have experimental data on 10,000–1,000,000 compounds. Searching for

*Computational Drug Design*. By David C. Young

literature results is an even more intensive task, as Chemical Abstracts Services adds over three-quarters of a million new compounds to its database annually.

Among other tasks, cheminformatics tools are used for curating data. More often than one would expect, it is necessary to integrate lists of compounds from new vendors, or those from newly acquired companies must be merged in with existing databases. Invariably, the structures are not normalized in the same way as the existing company resources. Normalization of structures may include adding hydrogens, setting protonation states, listing all tautomers, removing salt counter-ions, and specifying chirality. Some organizations will have a convention of storing structures in neutral form, whereas others will store the structures in the protonation state found at physiological pH. Software programs such as those sold by ACD/Labs can be used to predict $pK_a$ values and put compounds in the desired protonation state.

## 18.1 SMILES, SLN, AND OTHER CHEMICAL STRUCTURE REPRESENTATIONS

Large collections of data are kept in electronic form. This might be formatted as a relational database, spreadsheet, or flat file format. These records typically include a whole list of information about each compound, including the compound name, an internally assigned number, chemical formula, and experimental results. However, none of this information gives the chemical structure. The structure must be included in a format that allows researchers to search for a specific compound, similar compounds, and compounds with a specific structural motif. This is usually done by using a text string to represent the compound structure. These are called 2D structures, because they specify elements, connectivity, and bond order, but not the 3D Cartesian coordinates for each atom.

The two most commonly used 2D structure representations are SMILES and SLN. SMILES (Simplified Molecular Input Line Entry Specification) is most common, because Daylight Chemical Information Systems sells development libraries that allow chemistry software developers to easily integrate SMILES support into their applications. The advantage of SMILES is portability between applications. SLN (SYBYL Line Notation) is integrated into tools sold by Tripos, Inc. The advantage of SLN is that it provides a richer set of features for specifying molecular queries, stereochemistry, reactions, 3D coordinates, metadata, and more. Other molecular line notations include the Wiswesser Line Notation (WLN), ROSDAL (used by Beilstein), and InChI.

InChI (IUPAC International Chemical Identifier) was recently introduced by IUPAC as a standard for chemical structure representation. The primary

**TABLE 18.1 Examples of Chemical Structure Line Notation**

| Compound | Format | Line Notation |
|---|---|---|
| Ethanol | SMILES | CCO |
| Ethanol | SLN | CCOH |
| Ethanol | InChI | InChI=1/C2H6O/c1-2-3/h3H,2H2,1H3 |
| Ethylene | SMILES | C=C |
| Acetylene | SMILES | C#C |
| 2-Propanol | SMILES | CC(O)C |
| Benzene | SLN | C[1]H:CH:CH:CH:CH:@ |
| Benzene | SMILES | c1ccccc1 |
| Cyclohexane | SMILES | C1CCCCC1 |
| Phenyl search query | SLN | R1[hac>1]C[1]:C:C:C:C:C:@1 |
| Alanine with chirality | SLN | NH2C[S=SM]H(CH3)C(=O)OH<WEDGE_UP:=4, 6> |

reason for developing InChI is to give a unique identifier for an organic molecule. It has been extended to represent organometallic compounds as well. InChI is designed such that multiple tautomers will all have the same InChI string. InChI strings contain layers that are the main layer, charge layer, stereochemical layer, isotopic layer, fixed-H layer, and reconnected layer. The main layer is required and the others are optional. There is only one, unique, way to construct the InChI string for any given molecule or its tautomers. If two compounds have the same InChI main layer, they are identical, tautomers, or stereoisomers. InChI does not provide any advantages over other techniques for searching by similarity or substructure.

SMILES and SLN are very similar. In fact, many simple structures can be written so that they are valid under both formats. This is both due to the similarity and because both of these formats allow the same molecule to be written in several different ways. For example, the following are all valid representations of ethanol under either SMILES or SLN: CCOH, CH3-CH2-OH, C-C-OH. Most researchers store chemical data in files that utilize these formats, but do not look at the structure strings directly. However, the line notation specification determines what information can and cannot be stored in these files. Table 18.1 gives examples of these line notations.

## 18.2 SIMILARITY AND SUBSTRUCTURE SEARCHING

There are several ways in which one might wish to search for chemical structures. A researcher might be looking for data on a particular structure. However, there are also situations in which the researcher wants to carry out a substructure search to find a selection of compounds that all have the same structural motif but with different functional groups attached. There are also times when it is necessary to find compounds that are similar to a given compound, but do not have an identical structural motif. These types of searches are used to identify other compounds that should be screened once a few active compounds have been identified.

Substructure searches are usually pretty easy to perform. Most software packages that incorporate substructure searching have a graphical interface to build the search query. The software may also have an option to perform more complex searches by typing in a search query in the line notation used internally by that software. There are differences in functionality between software packages, due partly to the design of the graphical interface and partly to the abilities of the line notation system. As such, it may or may not be possible to perform more complex searches, such as specifying stereochemistry, lists of atoms that can appear at a given position, or more than one valid bond order.

The ability to find compounds that are similar to a known active compound is very useful in selecting compounds for screening. However, the notion of molecular similarity is based on the scientist's impression that one compound should behave similarly to another—not on a rigorous law of nature.

This concept of molecular similarity is implemented in computer programs as a distance, which is described in more detail below. The distance ranges from 0 to 1, with small values finding very similar compounds and large values finding very different compounds. Thus, finding compounds within a very small distance, say 0.02, might find only a few compounds that differ by one or two atoms or bonds. Many researchers use a distance of around 0.15 to find a group of reasonably similar compounds. Large distances between compounds ($>0.6$) give very dissimilar compounds. Searches with large distances are sometimes used to generate a diverse set of compounds. The gray area is in the intermediate range with a distance of 0.3–0.5. In this intermediate area, a compound that appears somewhat similar might have a larger distance than one that appears to bear little similarity. Some software packages invert this concept, thus defining a "similarity" that is 1 minus the "distance."

Most software packages implement molecular similarity using fixed length binary fingerprints. These fingerprints are a string of binary bits (1 or 0) that represent the presence or absence of various elements, functional groups, bond, rings, etc. A fingerprint is generated to numerically represent each compound, and then a distance formula is used to compute the distance between two compounds from their fingerprints. The most widely used distance formula is the Tanimoto distance.

The software documentation should be consulted when details of the fingerprint construction are needed. Here are comments on a few of the available fingerprint schemes. UNITY fingerprints (also called SLN fingerprints) are 988-bit or 992-bit, hashed, binary fingerprints. BCI (Barnard Chemical Information) fingerprints are 1052-bit binary strings. MACCS (also called MDL) fingerprints are binary bit strings that can be 164, 166, 320 or 960 bits long. Daylight fingerprints can be 2048-bit or 1024-bit binary strings.

Fingerprints can be hashed or unhashed. A hashed fingerprint uses a given bit for indicating the presence of more than one functional group. For example, this allows 5000 structural features to be represented by a 1024-bit fingerprint. The advantage of this is that hashed fingerprints can include more features, and thus give better performance on a wide range of chemical structures. Unhashed fingerprints can give better discrimination between very similar compounds in a very focused dataset. The majority of the fingerprint schemes in use are hashed fingerprints, as the advantages tend to outweigh the disadvantages. Hashed fingerprints perform surprisingly well, in spite of the obvious potential for error.

In general, binary, hashed fingerprints have proven to be a very effective tool for searching chemical databases, and for clustering together groups of compounds that have similar biological activity. However, there is always a worst case scenario. In the case of binary fingerprints, the worst case is comparing a compound with a dimer, trimer, etc. of the same structure. In this case, the binary fingerprints will show the compound, dimer, etc. to be extremely similar—more similar than one would expect based on a comparison of biological activity or normal boiling point.

There are several different ways of computing a distance from two fingerprints. One of the simplest is the Manhattan distance, which is a binary equivalent of taking a zig-zag path following ordinal axes only. The Manhattan distance is analogous to walking northeast in Manhattan, where it is only possible to take eastbound streets and northbound streets rather than following the diagonal. The Manhattan distance is sometimes used when a worst case estimate is desired. The Tanimoto distance is the most widely used distance formula for estimating a straight line diagonal path in binary fingerprint space. There are a dozen or so other fingerprint schemes that are used only on rare occasions. These include Cosine, Dice, Euclid, Simpson, and Yule.

This system of similarity distances is quantitative and useful, as long as one keeps in mind that it is a mathematical representation of the idea of a similarity, and not a law of nature. Graphical plots of similarity are nonlinear maps (Fig. 8.5) or other types of cluster representations. This is because the space of fingerprint distances is an unusual mathematical space. It is quantitative, but it is a nonlinear space that is non-Euclidean, and may not obey the triangle inequality. Thus, nonlinear maps are not a completely accurate representation of the space, but they are the best possible projection of this unusual space onto a two-dimensional Cartesian coordinate system.

Chemistry software also utilizes some other types of fingerprints that are not the same as binary fingerprints and similarity distances. For example, HQSAR (holographic quantitative structure activity relationship) uses a fingerprint that is not binary. HQSAR utilizes this fingerprint for creating structure–activity relationships, rather than for computing a similarity.

Binary fingerprints, such as UNITY and MACCS, are sometimes called 2D fingerprints to distinguish them from 3D fingerprints. Several vendors offer a 3D fingerprint scheme, which is dependent upon the compound shape. There are both rigid and flexible 3D fingerprint schemes. This is an intermediate step between 2D fingerprints and pharmacophore models. Although 3D fingerprints appear to be an elegant idea, studies have shown that they are actually worse than 2D fingerprints for clustering compounds in a way that separates active and inactive structures. This is apparently because 3D fingerprints depend upon the conformational shape of the molecule, but lose any dependence upon the chemistry of the compound.

Some other mechanisms have been explored for quantifying molecular similarity. The topological index is a number representing a metric derived from the bonding pattern within a molecule. It is a reasonable assumption that two compounds that are very similar should have topological indices that are close in value. In practice, this works well for a focused library, such as a set of compounds that are all derivatives of a single base structure. When more diverse collections of compounds are analyzed, topological indices perform more poorly than fingerprints. Very different classes of compounds can have very similar topological indices. Thus, similar compounds will have similar topological indices, but dissimilar compounds will not necessarily have dissimilar topological indices.

Another molecule description scheme that has been in development is the atom environment method. In atom environments, the relative position of functional groups is taken into account. Thus, atom environments see a carbonyl two atoms from a halide as being different from a carbonyl three atoms from a halide. These two scenarios would be seen as the same in binary fingerprint descriptions. Atom environments can give more accurate similarity searching than binary fingerprints, at the expense of somewhat longer execution times. In a large study by Bender et al., multiple databases of inhibitor assay results were searched to see if atom environments were better at finding the most active compounds when looking for compounds similar to one of the most active compounds. For some datasets, atom environments did as much as 10% better than binary fingerprints. For some datasets, the difference was negligible.

Similarity measures have also been based on the physicochemical properties of compounds. These methods may use a set of 1D QSAR descriptors, or a 3D molecular alignment. Such methods can find other compounds that have similar physical properties, without regard for whether the chemical structures bear any similarity.

A quantum mechanically based measure of similarity is provided by the COSMOsim program, which is an extension to COSMOfrag from COSMO logic. COSMOfrag has a large database of chemical fragments. Each fragment has results from a COSMO-RS (conductor-like screening model for real solvents) calculation. COSMO-RS is used to perform a quantum mechanical calculation to obtain the surface polarization of the molecular electron density in the presence of a solvent. These surface polarizations can be compared in order to identify molecules that are bioisosterically similar. This procedure is implemented with a database of stored surface polarizations in order to ake the process sufficiently to be used to compare large collections of molecules.

Another surface property-based similarity tool is LASSO from SimBioSys. This is similar to COSMOsim in that molecular surface properties are utilized. However, LASSO differs in that it forms these properties into a fingerprint of integer values. The advantage of using this fingerprint is that it is mostly independent of the molecule's conformational shape. This makes the analysis run faster, and it is not necessary to have a preprocess step putting all compounds into the lowest energy conformation, or to combine the analysis with a conformation search. The disadvantage is that the fingerprint does not distinguish where the various properties are on the molecular surface relative to one another; thus, some percentage of false positives is expected.

## 18.3 2D-TO-3D STRUCTURE GENERATION

Software packages built around 2D chemical structure representations can implement many features for sorting, searching, and analysis of chemical data. There are many chemical properties that can be computed from 2D structures, such as those most often utilized for 2D QSAR. However, there are some chemical properties that must be computed from 3D Cartesian coordinates. On a large database of potentially millions of compounds, it is not practical to utilize molecular mechanics conformation searches on every compound to generate reasonable 3D structures. Instead, knowledge-based conformation analysis programs, such as Concord and Corina, are used to quickly generate 3D coordinates that are very reasonable, although not perfectly optimized. Some line notations, such as SLN, allow the 3D coordinates to be optionally included in the 2D structure. Other software packages will store the 3D coordinates in a separate field of their spreadsheet or database.

In the vast majority of cases, these knowledge-based programs for generating 3D structures give the ground state conformer for organic compounds. Very flexible compounds can come out with a conformer that is near the lowest energy conformer. In rare cases, very unusual compounds may be generated with very unreasonable geometries. Some software packages have an option for running a molecular mechanics optimization, or one or two steps of that optimization, on the structures. The knowledge-based structure-generating algorithms are discussed in more detail in Chapter 17.

Once 3D atomic coordinates have been generated, they can be used to compute additional molecular properties, such as dipole moments and moments of inertia. In many cheminformatics programs, these values can be used for searching, sorting, selection, and sometimes for generating 2D QSAR models.

## 18.4 CLUSTERING ALGORITHMS

Clustering techniques are used to group together sets of similar compounds. Examination of these groups can be used to identify possible false-negative or false-positive screening results. It can also be used to identify other compounds that should be screened for activity. These applications of the clustering results will be discussed further in the next section. The following paragraphs discuss the various clustering algorithms that are available for use on chemical data.

There are two broad categories of clustering algorithms: hierarchical and nonhierarchical algorithms. Hierarchical algorithms generate multiple layers of groups that successively break down the set of compounds into a larger number of clusters, having fewer and fewer compounds in each cluster. The results of a hierarchical clustering algorithm are usually represented by a tree, or dendrogram (Fig. 8.4). The advantage of hierarchical clustering is that the researcher can easily select the degree of clustering desired. The disadvantage is that there may not be a clear metric for making an objective choice of the degree of clustering that should be utilized. Nonhierarchical clustering algorithms generate a single group of clusters, based on a similarity measure defined by the researcher. The following paragraphs discuss some specific algorithms for both types of clustering.

A hierarchical clustering algorithm can be defined based on the maximal common subgraph (MCS), also called the maximal common substructure. The MCS is the largest collection of atoms and bonds that are identical within an entire group of compounds. A hierarchy can be defined in which the top layers have large groups of compounds that may only have a carbon atom in common, while the lowest layers have many groups each containing a few compounds with a large conserved pattern of rings and chains. This is the clustering mechanism implemented in the Distill package from Tripos.

Clustering algorithms can also allow clusters to be overlapping or non-overlapping. Overlapping clusters allow a given compound to be in more than one cluster. Use of overlapping clusters is very rare in chemistry. The following paragraphs discuss the merits of a number of popular clustering algorithms.

The Jarvis–Patrick clustering algorithm creates a set of clusters, with each item appearing in only one cluster. It does this by defining a number of neighbors in common and a distance metric. If two points have the designated number of neighbors in common (meaning that they are within the specified distance), then the two points are in the same cluster. This is a deterministic algorithm; that is, the same inputs always give the same cluster members.

The *K*-means clustering algorithm generates a specified number *K* of clusters. This is done by randomly selecting *K* points, and then generating clusters from surrounding points by using the initial set of points as cluster centroids. Then, new centroids are determined, and the algorithm iterates until the centroids do not move. This gives a clustering that may change if done again, as the initial points are selected randomly, but typically changes very little if the input options are chosen sensibly. It also means that the results will differ depending upon how many clusters are specified. The fact that the algorithm uses a set number of clusters in place of a cutoff distance means that clusterings of two different datasets are not directly comparable.

Group average clustering (also called average-link clustering) is an iterative algorithm that combines clusters into larger clusters until some cluster separation criterion is met. The iterative process starts out with each point being considered a one-point cluster; then, in each iteration, the two clusters that are closest together are merged into one cluster. The "average" aspect (as opposed to single-link or complete-link clustering) is in the way in which clusters are compared with one another by comparing cluster averages or centroids. This results in a deterministic algorithm that ensures that all of the compounds in a cluster are similar to one another, and dissimilar from other clusters, with a reasonable time complexity. The end result of this process is a set of clusters with every point assigned to exactly one cluster, although there may be singleton clusters with only one point in them.

Ward's clustering algorithm is a technique for generating a hierarchical set of clusters, starting from all points being singleton clusters and working up to all points in the same cluster. Like group average clustering, Ward's clustering is an agglomerative algorithm; that is, it starts with individual points, and coalesces small clusters into larger clusters at every step. The metric for determining which clusters should be combined at each step is a computation of the heterogeneity of the resulting combined cluster. This heterogeneity value can be used to quantify how far down each cluster the split occurs. In other words, the distance from two clusters to the point where the two split apart is a quantitative measurement of the similarity between the two clusters.

The Guénoche algorithm also generates a cluster hierarchy. The metric for determining which clusters are to be combined is qualitative. Thus, it can indicate that cluster A is more similar to cluster B than to cluster C. It cannot indicate, for example, whether the similarity between clusters is slightly different or fourfold different.

Recursive partitioning is another algorithm for creating a hierarchical set of clusters. It is a divisive algorithm; that is, it starts with all compounds in one large cluster, and then splits clusters in two at each iteration. This is based on the artificial intelligence task of building a decision tree to predict some property by successively checking for criteria that will separate out groups of compounds. There are a number of variations on the recursive partitioning algorithm, which often use different criteria for dividing clusters. However, they all have in common that the best division is one that makes each cluster have all like compounds, and that division into two nearly equally sized clusters is better than splitting a singleton off from a large cluster.

Some programs generate clusters without using a generic clustering algorithm like those given above. The alternative is to have a predefined set of chemical families and simply categorize the set of compounds according to which family each family fits best. The advantage of this is that two completely different sets of compound structures can be categorized into clusters that are directly comparable. The disadvantage is that it relies upon the software developers and researchers to define a set of chemical families that will be reasonable for absolutely every possible compound.

Many older works utilized the Jarvis–Patrick clustering algorithm for nonhierarchical clustering. This was one of the earlier developed clustering algorithms, and it is easily implemented in code. Over the course of the cluster generation, the Jarvis–Patrick algorithm grows clusters by adding compounds at the edge of existing clusters. This can result in clusters that are long and thin in terms of clustering on chemical structure fragments, as shown in Fig. 18.1. For example, a cluster may have a series of compounds that differ by the length of a chain in the middle of the compound. Such a collection of compounds is certainly chemically related, but very nonuniform in terms of variance in molecular weight or conformance to a particular pharmacophore model. Thus, the Jarvis–Patrick algorithm has been reported to be one of the worst clustering techniques when tested for the ability to create clusters of mostly active or inactive drug candidates.

The Ward, Guénoche, *K*-means, and group average techniques all tend to perform well for clustering together groups of compounds in a way that reflects drug activity. In all cases, this is dependent upon the selection of appropriate clustering parameters. Often, the default values set by software packages will be the values most commonly giving the best results for drug activity analysis.

Clustering algorithms are also used for diversity selection. A clustering of a large number of compounds available in house or commercially can be done. A compound or several compounds can then be selected from each of the clusters, in order to get a diverse sampling of the entire chemical space. Such diverse libraries are often used in early rounds of compound screening.

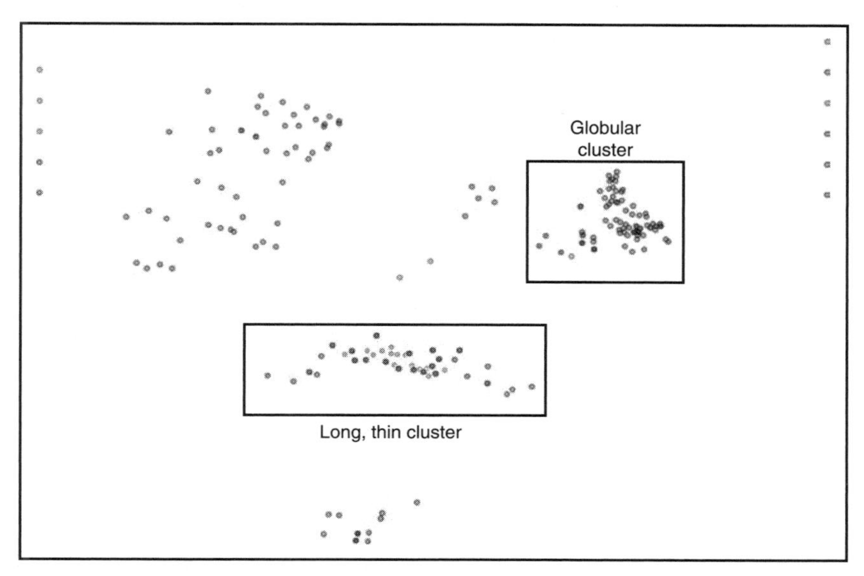


**Figure 18.1** A 2D nonlinear map illustrating the difference between long, thin clusters (generally not desired) and globular clusters. The compounds at the two ends of a long, thin cluster can be surprisingly dissimilar.

## 18.5 SCREENING RESULTS ANALYSIS

The preceding sections of this chapter have discussed various cheminformatics algorithms. This section will discuss one of the important uses of these techniques in the drug design process. Experimental screening results can be organized, analyzed, and used to make subsequent decisions about the next round of synthetic, experimental, and computational tasks to be undertaken. The old-fashioned way of doing things was to print all of the results on a sheet of paper and look at them carefully. That works well for projects where only some 20 compounds are synthesized and analyzed, which never occurs in the pharmaceutical industry today. It is now common to have anywhere from 10,000 to 1,000,000 structures screened, thus making it absolutely essential to have computer software for managing and analyzing all of the data. Software packages for performing these tasks are sometimes described as screening analysis packages, but may also be described more broadly as decision support tools.

The analysis of screening data is difficult, because there may be a very large number of data points, some of which can be false-positive or false-negative results. An additional complexity is the fact that researchers are trying to find compounds that meet a number of criteria, such as activity, bioavailability, and lack of toxicity. The reliability of the assay results is dependent upon the type of assay performed.

High throughput screening assays typically test each compound at a single concentration in a single well of a plate. With this approach, the largest number of compounds can be screened in the most cost-effective manner. Unfortunately, it also results in a high occurrence of false-positive and false-negative results. Compounds showing positive results are typically passed on to a more rigorous assay, regardless of whether computational analysis suggests that the results may be correct or incorrect.

False-negative results create greater difficulty, as a potentially valuable compound may be passed over and not tested again. There are several ways to minimize this risk. Some companies use low cutoff criteria in the high throughput assays, thus intentionally designing the assay to give a high percentage of false positives and minimize the number of false negatives. Another technique is to never submit singleton clusters to the assay, meaning that for every compound assayed there should be at least one other compound in the assay that is extremely similar. Thus, for every false negative, there should be a positive result, so that a structural motif will not be accidentally overlooked.

In the course of screening analysis, the chemist is trying to accomplish a number of tasks, including:

- identifying classes of compounds that are active against the biological target;
- identifying possible false-positive results;
- identifying possible false-negative results;
- evaluating the potential suitability of a compound based on criteria such as toxicity, pharmacokinetics, and sometimes even synthetic accessibility;
- identifying structure–activity relationships (SAR) that explain the activity of the positive compounds, and sometimes generating quantitative structure–activity relationship (QSAR) models.

Tools to search company structure databases and commercially available sources for other compounds that should be tested are sometimes integrated into screening analysis tools, and sometimes packaged separately. In rare cases, screening analysis software is integrated with software for designing new compounds and combinatorial libraries. However, this is most often done in another step with other tools.

The following paragraphs go through the process of screening results analysis. They are ordered based on the typical sequence in which these tasks might be performed. However, this process does vary somewhat from one company to the next, and even from one drug design project to the next.

The first step in screening results analysis is simply sorting the results in various ways. The compounds with the strongest activity typically receive particular attention. However, the most active compounds can be useless leads if they are particularly toxic, or if they are promiscuous compounds that bind to most proteins (see Fig. 2.4).

If there are a large number of active compounds, the researcher may use sorting in conjunction with binning and selection functions. Binning can be used to categorize compounds, for example ranking each compound on a scale of 1–5 based on whether it is predicted to be orally bioavailable. This type of data can in turn be used to select compounds that have high activity, are orally bioavailable, have low toxicity, etc.

Clustering is typically run on the entire collection of compounds assayed. Once this is complete, compounds that are very similar will be in the same cluster. The user can then select all of the active compounds, and extend that selection to all compounds that are in the same cluster as at least one active compound. This set of results is typically separated into a separate file and used in subsequent analysis steps, and the rest of the data is discarded. This typically results in a 90–99% reduction in the number of compounds being analyzed. This is still a conservative selection in that it errs on the side of keeping any compound that might possibly be useful.

The next step is typically to deal with the fact that early stage screening results are inexact. If the number of compounds kept from the previous step is not prohibitively large, all of them will be sent on to subsequent rounds of more accurate screening. Even if all of them are to be screened again, decisions for the next round of synthesis and high throughput assay must often be made before those screening results are complete. To facilitate these decisions, software analysis can be used to mark compounds as possible false positives and possible false negatives. When a few negative compounds are found in a cluster containing many positive compounds, they will be identified as possible false negatives. Of course, these negatives will be just as valuable if they are truly inactive, because those closely related inactives form the basis for identifying SAR. Likewise, a single active compound in a cluster containing many inactive compounds will be identified as a possible false positive.

At this point, predictive calculations are typically run. These might include predictions of compound toxicity, pharmacokinetic properties, and even synthetic accessibility. These types of calculations tend to also be of limited accuracy. One of the most difficult tasks for the researcher at this stage is to ignore any numbers in the second or third significant digit of most of the experimental and computational results.

The next step is to select a few chemical structural motifs to explore more thoroughly. Ideally, these should be motifs that appear in a number of active compounds. However, the cost of exploring a few dead ends at this stage of the design process is fairly low, so there is no harm in exploring an interesting singleton structure.

Once a structural motif, or several, has been selected for closer examination, the next step is to use searching tools to find any additional available compounds that can be assayed. This typically means searching through several databases of chemical structures: the database of compounds synthesized in house and those commercially available from a number of vendors. Substructure searches find compounds that have the exact pattern of atoms specified in the search query. This can find some very relevant compounds, but typically also finds some where that pattern of atoms appears in a compound that is far too large to fit in the target's active site. Similarity searching will find compounds that have a very similar size and shape, although not necessarily the exact same pattern of atoms.

SAR can be determined once reliable assay results are available for a reasonable number of compounds with the same structural motif. Generally, it is preferable to have 20 or more compounds assayed to construct SAR models—and a larger number is even better. Note that SAR is a qualitative model, meaning that it gives information such as the fact that a hydrogen bond donor should be attached to carbon 3. This is different from QSAR, which gives a quantitative, numerical prediction of some property, such as blood–brain barrier permeability.

One of the more sophisticated SAR techniques is structural unit analysis (SUA), which is part of the Benchware DataMiner package from Tripos, where it is named SAR Rules. Unlike other SAR techniques, SUA gives recommendations as to what type of structural group to use. It also can suggest structural sequences, such as group 1 connected to group 2 connected to group 3, as shown in Fig. 18.2.

Another SAR technique is to identify an activity cliff. This is a pair of very similar molecules with very different activities. If the pair of molecules differ by only one functional group, then the comparison gives information about what size and type (hydrogen bond donor, etc.) of functional group should be in this position. Figure 18.3 shows an activity cliff display from the Class Pharmer software from Simulations Plus, which refers to this as a pairSAR.

At some point, researchers will select one structural motif to develop, leaving the others as plan B or plan C options if the first one is not satisfactory. This is typically done before beginning traditional, organic synthesis work on a particular motif. It is sometimes done even earlier than that, before beginning combinatorial synthesis work, if enough data is available to reasonably make

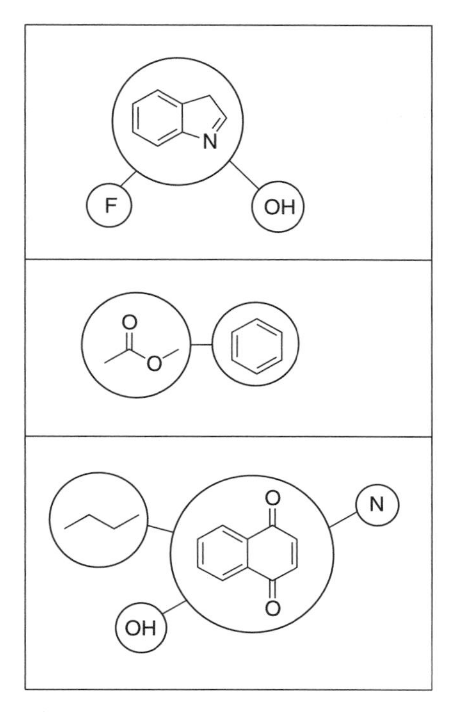


**Figure 18.2** Examples of the type of SAR rules that are generated by the structural unit analysis method.

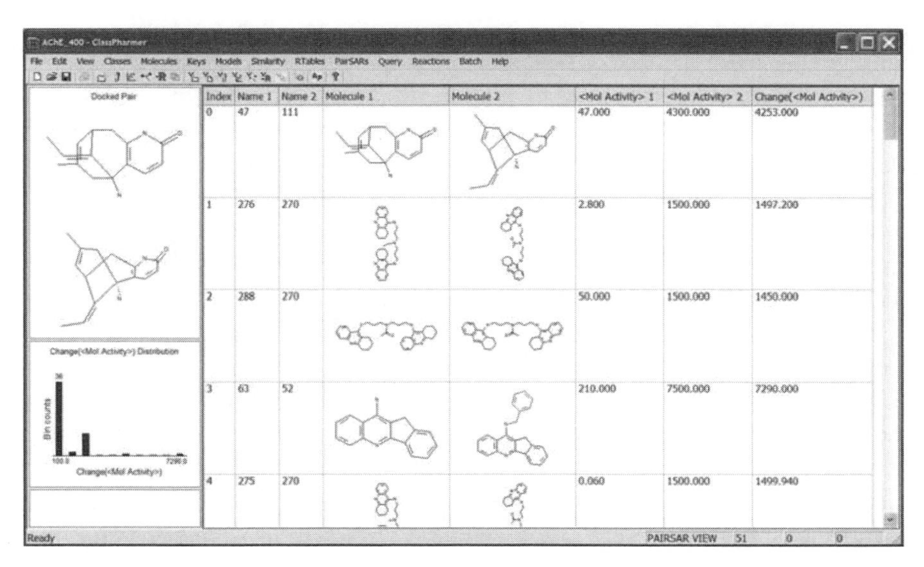


**Figure 18.3** The Class Pharmer pairSAR format for displaying an activity cliff.

that decision at that stage. At this point, researchers typically begin the process of structure-based drug design, testing docking experiments, and using many of the techniques discussed earlier in this book.

There are a variety of commercial software packages for performing analyses of high throughput screening data. Most are stand-alone programs designed specifically for chemistry applications, which typically interface with other chemistry applications at the very least through the import and export of shared data file formats. The exception to this format is Accord for Excel, a product sold by Accelrys. Accord is a plugin that adds chemical structure viewing and computations into Microsoft Excel. The advantage of this is that it leverages the feature set of Excel. The disadvantage is that it is limited by the limitations of Excel, such as the maximum number of rows that an Excel spreadsheet can contain. Packages specifically designed for handling large chemical datasets, such as Benchware HTS DataMiner from Tripos and Class Pharmer from Simulations Plus, can handle much larger datasets.

Most high throughput screening analysis tools include clustering algorithms and various ways of displaying chemical data. These are the tools needed for a scientist to find the correct data in order to manually develop an SAR. A few software packages include automated algorithms, such as structural unit analysis, to automate the generation of SAR rules. Many researchers use more than one software package, depending upon the needs of a given project.

## 18.6 DATABASE SYSTEMS

This chapter has discussed large collections of data, such as chemical structures, experimental results, and molecular properties. These may be compounds on hand, compounds available commercially, or compounds that have only been designed *in silico* but have not yet been synthesized. Regardless of the origin of the data, much of the software to work with that data is a database of some sort. The software packages for accessing these databases seem to follow one of two paradigms: some are designed to be used by a single researcher; others are designed to be enterprise information systems, accessible to a large number of employees.

Individual user database systems are usually self-contained. However, in some cases, it is necessary to also install some non-chemistry-specific database software in order to use them. The advantage of individual user systems is that the user has control over how the compounds are stored, imported, exported, etc. The disadvantage is that at least the graphical front end is often limited by the amount of memory available on a desktop computer. This can limit the number of compounds that can be worked with to a few tens of thousands. However, there are systems that keep most of the data on the disk drive, thus enabling them to work with millions of compounds on an average desktop computer. When purchasing such systems, it is important to be aware that even a system that can work with incredibly large collections of data may have certain features that can only be used on a smaller subset of the data.

One product that is different from many of the others is UNITY from Tripos. UNITY is a collection of tools that work on Tripos databases, which are flat files containing a set of SLN strings. The advantage of this is that it is very easy for a researcher to set up a Unix script to call the UNITY tools and feed data through. This ability to be readily customizable and automatable can be very valuable when handling large amounts of data.

Enterprise information systems tend to be large, complex systems, which have been customized for the needs of each company. Most of them contain a large amount of functionality to search for compounds by name, structure, company numbering system, etc. These systems may be interconnected with systems for purchasing, record keeping, inventory, etc. If such a system is integrated into an electronic laboratory notebook system, there may be data fields that are write-only and some form of electronic signature.

Enterprise information systems are usually heavily customized, often pieced together from systems from different vendors and some code written just for a particular company. At one time, MDL (now a division of Symyx) was the dominant provider of large chemical databases. In recent years, products from Tripos, Accelrys, and other companies have been gaining market share. Many pharmaceuticals have been making significant changes to part or all of their large data systems owing to the electronic record keeping regulations being adopted by the FDA in order to conform to federal regulations prescribed by 21 CFR Part 11. Another current trend is the integration of tools with online, web-based data repositories.

Cheminformatics software is not a high profile, core prediction technique. None the less, it is a necessary part of the drug design and development process, both computational and experimental.

## BIBLIOGRAPHY

### Books

Bajorath J, editor. Chemoinformatics; Concepts, Methods, and Tools for Drug Discovery. Totowa, NJ: Humana Press; 2004.

Daylight Theory Manual. Daylight Version 4.9. Aliso Viejo, CA: Daylight Chemical Information Systems; 2008. Available at http://www.daylight.com/dayhtml/doc/theory/index.html. Accessed 2008 Oct 21.

Ekins S, Wang B, editors. Computer Applications in Pharmaceutical Research and Development. Hoboken, NJ: Wiley-Interscience; 2006.

Everitt BS. Cluster Analysis. London: Edward Arnold; 1993.

Leach AR, Gillet VJ. An Introduction to Chemoinformatics. Dordrecht: Springer; 2007.

Gasteiger J, Engel T, editors. Chemoinformatics. Hoboken, NJ: Wiley; 2003.

Oprea TI, Mannhold R, Kubinyi H, Folkers G, editors. Chemoinformatics in Drug Discovery. Weinheim: Wiley-VCH; 2005.

## Fingerprint Similarity Review of Comparison Studies

Bender A, Mussa HY, Glen RC, Reiling S. Similarity searching of chemical databases using atom environment descriptors (MOLPRINT 2D): Evaluation of performance. J Chem Inf Comput Sci 2004; 44: 1708–1718.

Downs GM, Willett P. Similarity searching in databases of chemical structures. Rev Comput Chem 1996; 7: 1–66.

Willett P, Barnard JM, Downs GM. Chemical similarity searching. J Chem Inf Comput Sci 1998; 38: 983–996.

Willett P. Structural similarity measures for database searching. In: von Ragué Schleyer P et al., editors. Encyclopedia of Computational Chemistry. New York: Wiley; 1998. p. 2748–2756.

## Clustering Review or Comparison Articles

Barnard JM, Downs GM. Clustering of chemical structures on the basis of 2-D similarity measures. J Chem Inf Comput Sci 1992; 36: 644–649.

Brown RD, Martin YC. Use of structure–activity data to compare structure-based clustering methods and descriptors for use in compound selection. J Chem Inf Comput Sci 1996; 36: 572–584.

Downs GM, Willett P. Clustering of chemical structure databases for compound selection. In: van de Waterbeemd H, editor. Advanced Computer-Assisted Techniques in Drug Discovery (Methods and Principles in Medicinal Chemistry. Volume 3). Weinheim: VCH; 1994.

Additional references are contained on the accompanying CD.

# 19

# ADMET

ADMET stands for "adsorption, distribution, metabolization, excretion, and toxicity." These cover the pharmacokinetic issues determining whether a drug molecule will get to the target protein in the body, and how long it will stay in the bloodstream before another dose must be administered. Toxicity is often grouped with the pharmacokinetic properties, because similar software tools are used to predict the toxicity and computational toxicity predictions are usually performed at the same point in the design process as computational ADME calculations.

Recent years have seen a rise in the importance of computational ADMET predictions. This is because the majority of clinical trial failures have been due to ADMET issues, not from a lack of efficacy. Since this is the most costly point to have a failure, ADMET-related research and development activities can pay for themselves for decades to come if they can divert even one clinical trial failure.

The prediction of ADMET properties is incredibly very difficult task. As a compound moves from the digestive tract to the drug target, it will encounter tens of thousands of proteins, lipids, blood borne compounds, etc. The present state of technology is not up to the task of predicting every potential interaction and the probability of that interaction occurring. Thus, most ADMET predictions are made on the basis of relatively simple data, such as compound size, lipophilicity, and functional groups present. This type of information is used to

*Computational Drug Design*. By David C. Young

develop a QSAR model, group additivity model, or neural network. There are QSAR programs designed specifically for creating ADMET prediction models, such as ADMET Predictor from Simulations Plus.

Most of the ADMET QSAR equations are developed using the standard practice of finding correlations between the property to predict and descriptor values. One program that takes a somewhat different approach is COSMOtherm from COSMOlogic. This program starts with a quantum mechanical calculation that includes solvation effects via the COSMO (conductor-like screening model) continuum solvent method. It then creates a surface polarization function using a process called COSMO-RS (COSMO for real solvents). That function is used in a statistical mechanical process of creating distribution functions and integrating over them to obtain chemical potential and activity coefficients. The ADMET predictions produced by COSMOtherm, such as blood–brain barrier permeability, are generated from a linear combination of the terms that went into the chemical potential calculation. This final step is a QSAR model, but it is based on the type of solvated

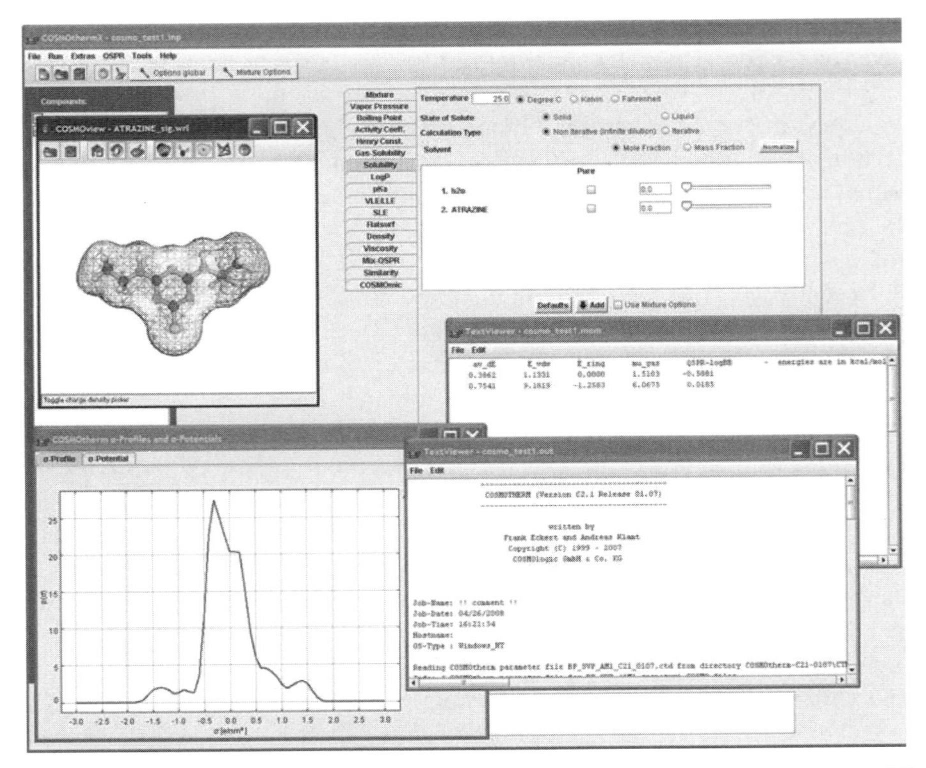


**Figure 19.1** A sample display from the COSMOthermX program, which computes ADMET properties based on quantum mechanical and statistical mechanical calculations.

molecular properties that govern the natural process, at least for passive diffusion. Figure 19.1 shows a screen shot of COSMOthermX, the graphical interface for COSMOtherm.

Owing to the difficulty of finding an efficient manner to predict the behavior of a very complex system, ADMET prediction tools tend to be only 60–70% accurate, meaning that the results are only qualitatively correct about 65% of the time. Some toxicity models do somewhat better, as they are developed for one specific type of toxicity.

Some pharmaceutical groups purchase all of the major ADMET prediction tools on the market. They then make predictions with all of the tools and look for compounds where the multiple tools give a consensus as to behavior. This consensus scoring technique tends to give results that are qualitatively correct up to 85% of the time.

Because computational predictions are so crude, drug design researchers tend to use the computations as a "red flag" to indicate a potential risk. They will also use ADMET calculations as a tie breaking vote when trying to make decisions between two, otherwise comparable, research options.

## 19.1 ORAL BIOAVAILABILITY

It is preferable to have drugs that can be administered orally in tablet or similar form, which is referred to as being orally bioavailable. If the drug is not orally bioavailable, then it must be administered by injection. In order for a drug to be orally bioavailable, the drug must be adsorbed from the digestive tract into the bloodstream. In some cases, it may be necessary for the drug to get through the cell membrane.

The majority of orally bioavailable drugs enter the bloodstream through passive intestinal absorption in the small intestine. Less common mechanisms for entering the bloodstream are through active transport in the small intestine, absorption in the stomach, absorption in the large intestine, nasal uptake, and topical absorption. Thus, one design criterion is that the drug should be soluble at the pH of the region of the digestive tract in which it is to be absorbed into the bloodstream, as shown in Table 19.1.

Active transport utilizes the body's mechanisms that force nutrient or signaling molecules to pass through a barrier. It is a very efficient mechanism to get a drug to its target. Unfortunately, it is often not possible to make a drug molecule that has the desired efficacy while simultaneously efficiently fitting the profile of an active transport substrate.

One of the active transport systems that does improve the uptake of a non-trivial percentage of drugs is the human intestinal small peptide carrier (hPEPT1). hPEPT1 is a low affinity, proton-coupled, active, oligopeptide

**TABLE 19.1 Digestive System Average Physiological pH**

| Location | Fasted | Fed |
|---|---|---|
| Stomach | 1.3 | 4.9 |
| Duodenum | 6.0 | 5.4 |
| Jejunum | 6.3 | 5.7 |
| Ileum | 7.0 | 7.0 |
| Caecum | 6.4 | 6.4 |
| Colon | 6.8 | 6.8 |

transporter. It has a broad substrate specificity that enables transport of the di- and tripeptides occurring in food products. hPEPT1 is active in the transport of a range of pharmaceuticals, such as β-lactam antibiotics and angiotensin-converting enzyme (ACE) inhibitors. Depending upon the nature of the drug, vitamin transporters, the glucose transporter (GLUT2), and nucleoside transporters may also play a role in drug transport.

One effective way to use active transport mechanism is to create a prodrug. A prodrug is a molecule that can be viewed as two compounds, one that fits the active transport profile and one that is the active drug, interconnected by a group that will be cleaved by one of the native enzymes. Thus, the active transport substrate moves the prodrug to the correct location, and then an enzyme cleaves it to release the drug molecule. Prodrug approaches are discussed further in Chapter 26.

Passive adsorption is simply a matter of giving the compound physical properties that allow it to get to the correct location. This typically means making the compound lipophilic enough to pass into and then through the cell membrane. The rule of thumb in drug design of passively adsorbed compounds is "*Make it just lipophilic enough to get to the target.*" It is not good to make the drug any more lipophilic than necessary, because very lipophilic compounds are readily removed from the bloodstream by the liver. If the drug is removed by the liver too readily, then it is necessary to increase the dosage frequency.

Most computational tools for predicting ADMET properties are some form of QSAR model. These might be linear QSAR equations, nonlinear equations, or group additivity models. The most common tools of this type are passive intestinal absorption models (almost always absorption in the small intestine) and models to predict Caco-2 colon wall permeability. Caco-2 is used because it is one of the few experimental bioavailability measurements available. It is hoped that other ADME properties, such as cell membrane (and perhaps cell wall) permeability will be proportional to the Caco-2 results. Also in use are more physical chemistry-based predictions, such as solubility, log $P$, and log $D$.

## 19.2 DRUG HALF-LIFE IN THE BLOODSTREAM

When a drug is administered, its concentration in the blood initially rises rapidly to some peak value as it is adsorbed into the bloodstream, and then slowly falls as it is eliminated from the body. The half-life is the amount of time required for the concentration in the bloodstream to be reduced to half of the peak concentration. Depending upon the rates of adsorption and elimination, the drug may have a half-life in the bloodstream anywhere from half a day to 5 days. The half-life in the bloodstream can be affected by the patient's diet, whether the drug is taken with food, the acidity of the drug, the formulation of the drug's oral form, and whether other drugs are being taken concurrently.

Most of the computational ADMET tools on the market simply predict one specific type of adsorption, toxicity, etc. However, it is possible to combine those models to look at the rate of drug entering the body and the rate of elimination to arrive at a prediction of the half-life in the bloodstream. This is, in turn, an indication of the necessary dosage frequency. Most of these predictions are made with some sort of QSAR model. There have been studies using techniques such as molecular dynamics to examine processes in membranes, but these techniques are not yet sufficiently developed, or fast enough, to render an accurate answer quickly and without user interactions.

The primary mechanism for removing drugs from the body is breakdown by cytochrome P450 (CYP) enzymes in the liver. Studies have indicated that human liver CYP3A4 is responsible for the metabolism of about 50% of all drugs currently in use. Six human CYP enzymes—1A2, 2C9, 2C19, 2D6, 2E1, and 3A4—together account for the oxidation of more than 90% of the drugs in use for human treatment. Some have a pronounced tendency to metabolize a given class of drugs; for example, CYP2C19 and CYP2D6 are responsible for the metabolism of many antidepressants. Many other CYP enzymes have also been indicated in the metabolism of drugs.

If a drug is also a CYP inhibitor, then administration of that drug will slow the removal of other drugs taken concurrently from the bloodstream. As a result, precautions must be taken to adjust dosages of drugs that are taken along with CYP inhibitors. Because of this role that CYP enzymes play in drug–drug interactions, there has been interest in predicting whether compounds will be CYP inhibitors. Note that this can be somewhat more complex than simply running a docking simulation of the drug in the CYP active site. Docking calculations are designed for predicting interactions with an enzyme that natively binds very specifically to one specific substrate. CYP enzymes are unusual in that they natively bind and react with broad classes of compounds. There are additional references on the topic in the supplemental references file on the CD accompanying this book.

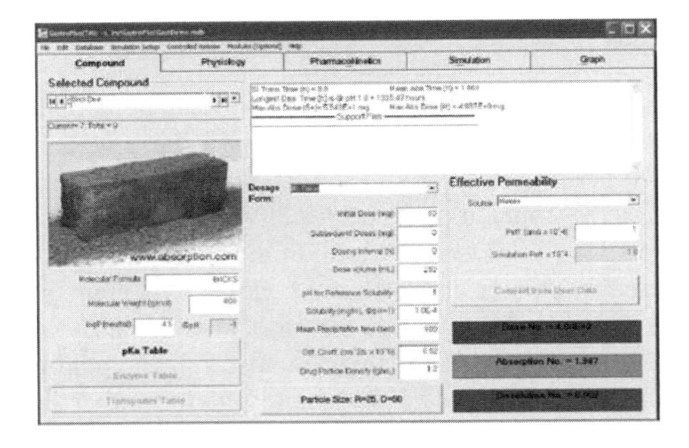


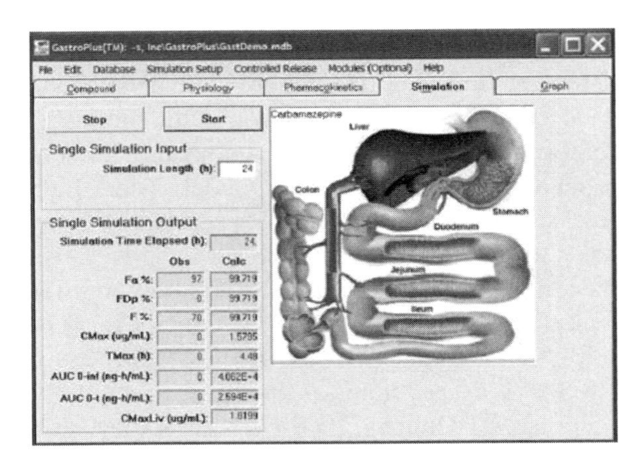


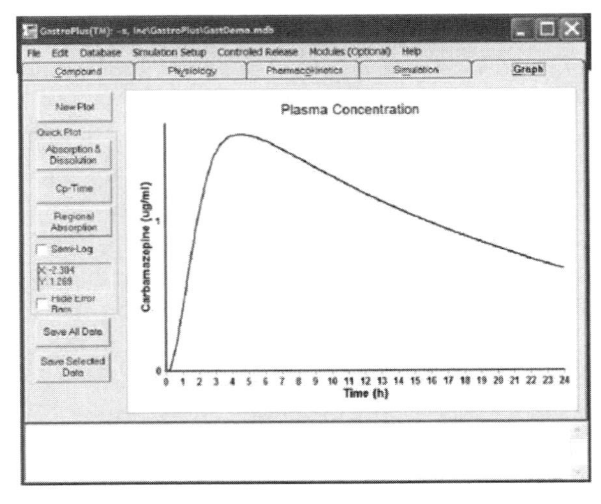


**Figure 19.2** Screen shots from Gastro Plus, sold by Simulations Plus.

Estimating drug elimination from the body can be as simple as a lipophilicity prediction. More often, this is done by use of a QSAR model parameterized to estimate interaction with CYP enzymes in general, but not any one specific enzyme. The program isoCYP from Molecular Networks predicts which CYP enzyme is most likely to be the one metabolizing a given small molecule. Having this type of tool available could lead to having separate models for each CYP in the next generation of software packages.

The leading software package for combining multiple rate equations into a half-life calculation is the GastroPlus software from Simulations Plus (which makes several different ADMET-related software packages). This takes into account the physical properties of the drug. Figure 19.2 shows examples of the GastroPlus graphical interface. At present, corrections for patient diet and other drugs taken concurrently are beyond the capabilities of any prediction software.

## 19.3 BLOOD–BRAIN BARRIER PERMEABILITY

The blood–brain barrier (BBB)is a film of epithelial cells that surrounds the brain and the rest of the central nervous system (CNS). It prevents the majority of blood-borne compounds from reaching these highly sensitive organs. In general, drugs that must reach targets within the CNS must be lower in molecular weight than other drugs and fall within a narrow lipophilicity range. The majority of CNS-active drugs have a log $P$ value of about 2. These requirements are based on passive transport through the BBB, which is what all of the available computational prediction tools model. There are also prodrugs that cross the BBB via an active transport mechanism.

A number of software vendors sell BBB prediction tools, which are typically forms of QSAR models. There is generally not a large difference in the accuracy of prediction from one BBB prediction tool to the next.

## 19.4 TOXICITY

Toxicity can also be a source of clinical trial failures, or a reason for drugs being removed from the market. Some of the drugs that have been removed from the market due to toxicity concerns are azaribine, nomifensine, encainide, temafloxacin, flosequinan, mibefradil, astemizole, troglitazone, rofecoxib, ximelagatran, gatifloxacin, and Fen-Phen. Six of these drugs were removed from the market because of cardiotoxicity concerns.

Computational prediction of toxicity has both good and bad points, compared with prediction of ADME properties. The good side is that toxicity prediction tends to be somewhat more accurate than ADME prediction. This is because a toxicity prediction model will predict results for one, specific type of toxicity. The downside is that there are many types of toxicity, and only prediction models for a handful of these situations. QSAR models exist for mutagenicity, carcinogenicity, teratogenicity, aquatic toxicity, hepatotoxicity, biodegradation, bioconcentration, bioaccumulation, maximum tolerated dose, acute toxicity, and cardiotoxicity (often hERG binding). For each type of toxicity, it may be necessary to have different computational techniques for various dose responses, such as acute dosage or chronic dosage.

Most toxicity prediction is done with the same sort of computational algorithms that are used for other ADME properties, such as QSAR algorithms and expert systems. Some researchers have been working on developing an innovative algorithm called an inverse docking algorithm. Unlike docking, which tests many ligands against one protein, inverse docking tests a single ligand against many protein targets. In principle, inverse docking should be an excellent way to do a toxicity prediction. In practice, it may be some years before this becomes the method of choice. Some problems with the development of inverse docking are that it is time-consuming and requires highly automated algorithms, and that accurate structures are not available for many biological targets.

Liver toxicity (hepatotoxicity) is a side effect of many drugs. As a consequence, a large percentage of the aging population in the United States has some degree of liver damage, and the trend is increasing. In response to this, the FDA tends to increase acceptable safety limits as new drugs are developed to replace older treatments. Hepatotoxicity is thus an increasing concern in drug design.

In recent years, there has been a surge in interest in predicting whether drugs will inadvertently block the potassium ion channels created by the hERG (human ether-a-go-go related gene) protein. Such blockage can result in an unwanted side effect of drug-induced cardiac arrhythmia. This tends to be caught late in clinical trials or, worse yet, after the drug has gone to market. Between 1998 and 2002, cardiac side effects were the cause of half of the drug withdrawals from the market. Thus, detecting such problems early can avert a very costly failure. A number of research groups have published QSAR or knowledge-based models for predicting hERG channel blocking (see the supplemental references file on the CD accompanying this book). The first commercially available flexible docking analysis has recently been announced by Quantum Pharmaceuticals.

Several other types of toxicity are routinely checked through the use of computer models. Carcinogenicity models are used to check for severe safety issues. Models of rat oral $LD_{50}$, the amount taken orally that results in a 50% death rate of the test animals, are used as a general measure of toxicity. The computer models predict toxicity to rats simply because all of the data used to develop those models is from rat studies.

Mutagenicity prediction is also utilized in the pharmaceutical industry. This software is available because there is sufficient Ames mutagenicity data to generate predictive computer models. The reason for doing this is that a very mutagenic compound may also exhibit unacceptable teratogenicity, causing fetal malformations if taken during pregnancy. It could also be an indication that the drug should not be administered to infants or children.

There are a number of toxicity prediction software packages on the market. Some of the toxicity prediction programs that have been established leaders in

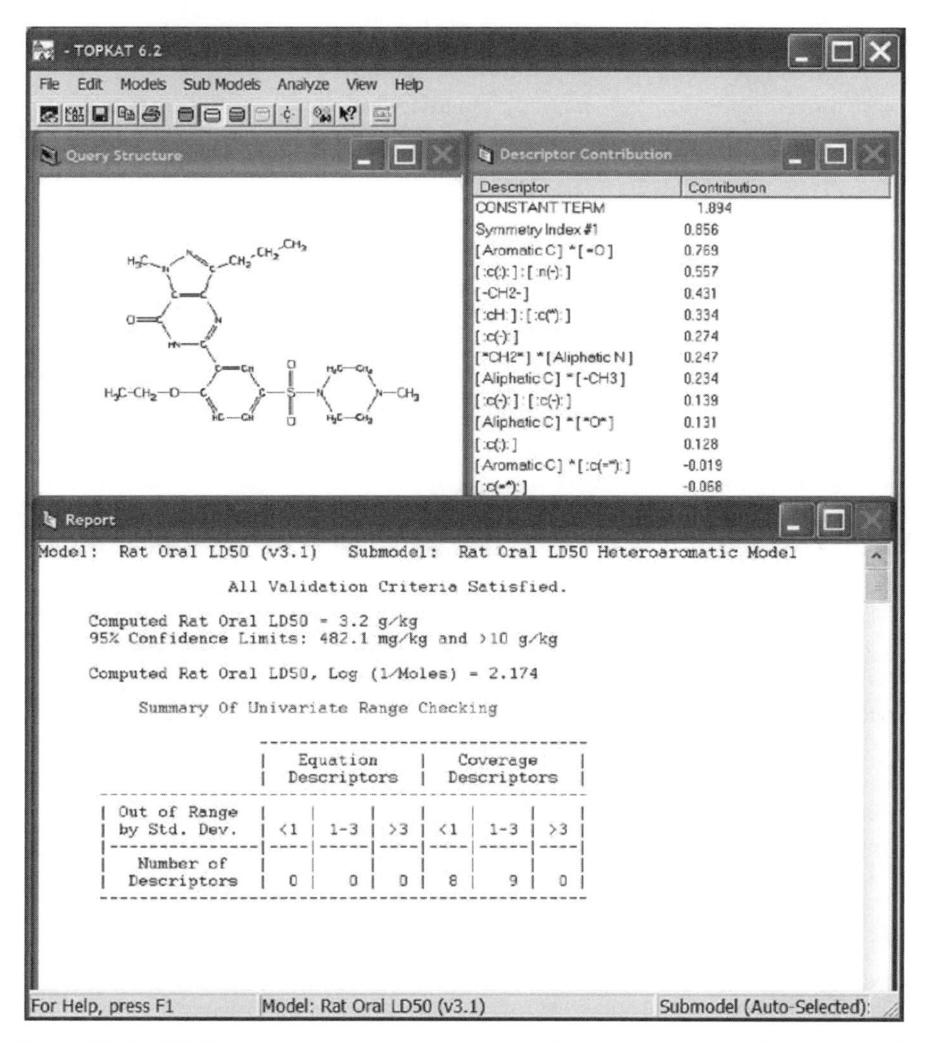


**Figure 19.3** TOPKAT from Accelrys is a toxicity prediction program that gives an analysis of how it arrived at the result, and an estimation of the accuracy of results.

the industry for some years are MCASE, Topkat, and DEREK. MCASE (Multi-Computer Automated Structure Evaluation) from Multicase Inc. is a software package that utilizes a large database of known toxicities and a collection of QSAR models to give a toxicity prediction. Topkat from Accelrys uses a group additivity method. One advantage of Topkat is that it gives estimated error bounds to provide some indication of the accuracy of prediction, as shown in Fig. 19.3. DEREK (Deductive Estimation of Risk from Existing Knowledge) is an expert system for giving a more qualitative estimation of toxicity. The q-Tox software, developed at Quantum Pharmaceuticals, is a newer toxicity prediction package, which uses an expert system approach. Other toxicity prediction programs are LAZAR, ToxScope, Hazard Expert, Tox Boxes, and OncoLogic.

In spite of their limited accuracy, ADMET prediction techniques are often utilized for every compound synthesized, and often every compound designed computationally. Even a remote possibility of preventing a failed clinical trial is worth the effort.

## BIBLIOGRAPHY

### Books

Dermietzel R, Spray DC, Nedergaard M, editors. Blood–Brain Barriers: From Ontogeny to Artificial Interfaces. Weinheim: Wiley-VCH; 2006.

Ekins S, editor. Computational Toxicology: Risk Assessment for Pharmaceutical and Environmental Chemicals. Hoboken, NJ: Wiley; 2007.

Klaassen CD, editor. Casarett & Doull's Toxicology: The Basic Science of Poisons. 6th ed. New York: McGraw-Hill; 2001.

Mannhold R, Kubinyi H, Folkers G, editors. Molecular Drug Properties: Measurement and Prediction (Methods and Principles in Medicinal Chemistry). Weinheim: Wiley-VCH; 2007.

Smith DA, van de Waterbeemd H. Pharmacokinetics and Metabolism in Drug Design. Weinheim: Wiley-VCH; 2006.

van de Waterbeemd H, Lennernäs H, Artursson P, Mannhold R, Kubinyi H, Folkers G, editors. Drug Bioavailability: Estimation of Solubility, Permeability, Absorption and Bioavailability. Weinheim: Wiley-VCH; 2003.

### Cost of Clinical Trial Failures Due to ADMET

DiMasi JA. Risks in new drug development: approval success rates for investigational drugs. Clin Pharmacol Ther 2001; 69: 297–307.

Kennedy T. Managing the drug discovery/development interface. Drug Discov Today 1997; 2: 436–444.

## Reviews of ADMET Prediction Techniques

Ajay. Predicting drug-likeness: Why and how? Curr Top Med Chem 2002; 2: 1273–1286.

Baringhaus KH, Matter H. Efficient strategies for lead optimization by simultaneously addressing affinity, selectivity and pharmacokinetic parameters. In: Oprea TI, editor. Chemoinformatics in Drug Discovery. Weinheim; Wiley-VCH: 2000, p. 333–379.

Ekins S. Swaan PW. Development of computational models for enzymes, transporters, channels, and receptors relevant to ADME/Tox. Rev Comput Chem 2004; 20: 333–415.

Feng J, Lurati L, Ouyang H, Robinson T, Wang Y, Yuan S, Young SS. Predictive toxicology: Benchmarking molecular descriptors and statistical methods. J Chem Inf Comput Sci 2003; 43: 1463–1470.

Gopi Mohan G, Gandhi T, Garg D, Shinde R. Computer-assisted methods in chemical toxicity prediction. Mini Rev Med Chem 2007; 7: 499–507.

Lewis DFV. Computer-assisted methods in the evaluation of chemical toxicity. Rev Comput Chem 1992; 3: 173–222.

Paranjpe PV, Grass GM, Sinko PJ. *In silico* tools for drug absorption prediction: experience to date. Am J Drug Deliv 2003; 1: 133–148.

Refsgaard HHF, Jensen BF, Brockhoff PB, Padkjær SB, Guldbrandt M, Christensen MS. *In silico* prediction of membrane permeability from calculated molecular parameters. J Med Chem 2005; 48: 805–811.

Selassie CD, Garg R, Kapur S, Kurup A, Verma RP, Mekapati SB, Hansch C. Comparative QSAR and the radical toxicity of various functional groups. Chem Rev 2002; 102: 2585–2605.

## Cytochrome P450-Specific Information

Flockhart DA. Drug Interactions: Cytochrome P450 Drug Interaction Table. Indiana University School of Medicine; 2007. Available at http://www.medicine.iupui.edu/flockhart/table.htm. Accessed 2008 Oct 20.

Additional references are contained on the accompanying CD.

# 20

# MULTIOBJECTIVE OPTIMIZATION

Drug design is not simply the task of making a compound good for one thing. It is the task of making a compound simultaneously good for several things. The drug should be active, nontoxic, orally bioavailable, free from side effects, etc., all at once. Multiobjective optimization (also called "multidimensional optimization") is the process of finding solutions that satisfy multiple criteria at the same time. Multiobjective optimization is a relative newcomer to the computational drug design field. However, it fills such a valuable role that it is this author's opinion that multiobjective optimization (or an algorithm filling the same need) will soon become part of the mainstream of computational drug design.

The way in which drugs are designed to meet multiple criteria has been changing over time. The traditional approach that has been followed over the past few decades has been to design drugs for improved activity first, and then to address other issues. This is not an unjustified approach, as it is generally easier to fix an ADMET problem than to improve activity. Many drug design organizations, therefore, still hold primarily to this manner of thinking. However, the downside is that optimizing activity first sometimes results in spending an excessively long time on the development of a dead-end lead.

The second step towards multiobjective optimization is to develop drugs for activity, and then to use pharmacokinetics as the tie breaker between multiple

*Computational Drug Design*. By David C. Young

leads. This adds ADMET issues into the drug design process, albeit at one discrete step in the process.

The third iteration in the evolution of drug design processes to utilize a greater number of drug design criteria has been to test *in silico* models for all of the criteria in parallel. Typically, all available computer models will be used, and a few of the most critical criteria will be checked experimentally. This gives researchers all of the correct data, even if less accurate than desired, but not a way to make sense of the onerous mass of data. This is the current state of affairs in many drug design laboratories.

The most recent step in this changing drug design process has been to utilize software tools designed for multiobjective optimization. A few such programs exist, utilizing significantly different algorithms. Figure 20.1 shows an

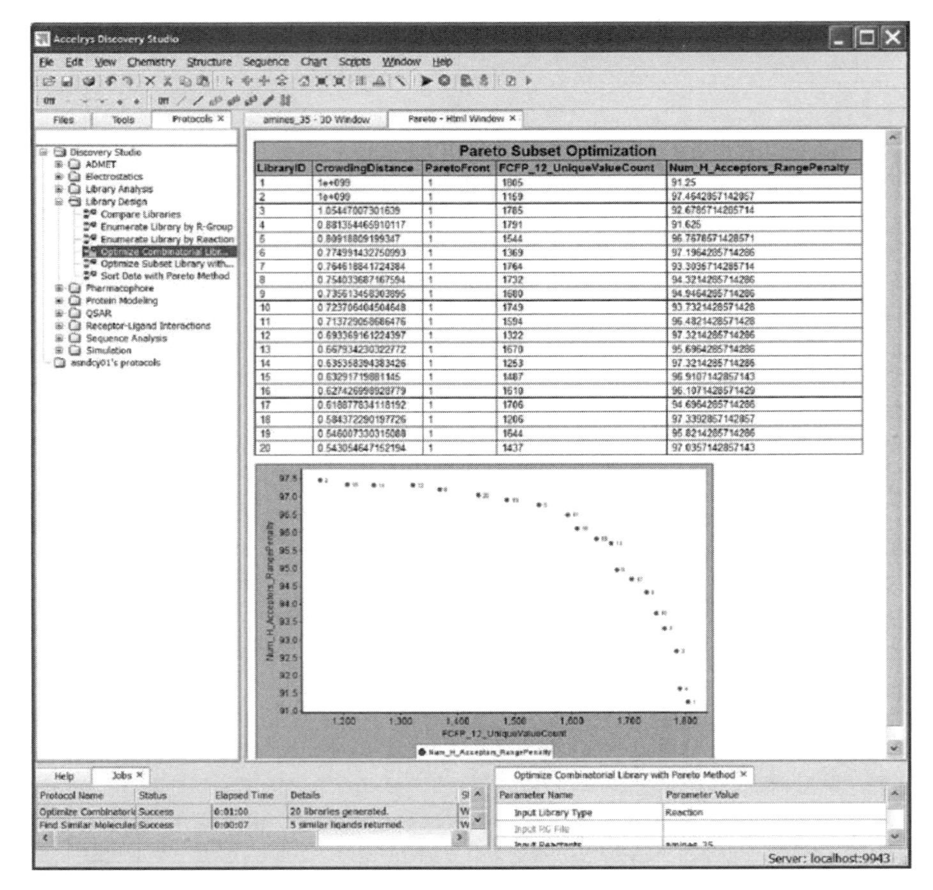


| LibraryID | CrowdingDistance | ParetoFront | FCFP_12_UniqueValueCount | Num_H_Acceptors_RangePenalty |
|---|---|---|---|---|
| 1 | 1e+099 | 1 | 1805 | 91.25 |
| 2 | 1e+099 | 1 | 1159 | 97.4642857142857 |
| 3 | 1.05447007301639 | 1 | 1785 | 92.6785714285714 |
| 4 | 0.881354465910117 | 1 | 1791 | 91.625 |
| 5 | 0.809188809199347 | 1 | 1544 | 96.7678571428571 |
| 6 | 0.774991432750993 | 1 | 1369 | 97.1964285714286 |
| 7 | 0.764618841724384 | 1 | 1764 | 93.3035714285714 |
| 8 | 0.754033687167594 | 1 | 1732 | 94.3214285714286 |
| 9 | 0.735613458303896 | 1 | 1680 | 94.9464285714286 |
| 10 | 0.723706404504648 | 1 | 1749 | 93.7321428571428 |
| 11 | 0.713729068686476 | 1 | 1594 | 96.4821428571428 |
| 12 | 0.693369161224397 | 1 | 1322 | 97.3214285714286 |
| 13 | 0.667934230322772 | 1 | 1670 | 95.6964285714286 |
| 14 | 0.635358394383426 | 1 | 1253 | 97.3214285714286 |
| 15 | 0.632917159881145 | 1 | 1487 | 96.9107142857143 |
| 16 | 0.627426998928779 | 1 | 1610 | 96.1071428571429 |
| 17 | 0.618877834118192 | 1 | 1706 | 94.6964285714286 |
| 18 | 0.584372290197726 | 1 | 1206 | 97.3392857142857 |
| 19 | 0.546007330315088 | 1 | 1644 | 95.8214285714286 |
| 20 | 0.543054647152194 | 1 | 1437 | 97.0357142857143 |

**Figure 20.1** The results of a Pareto multiobjective optimization as displayed in the Discovery Studio program from Accelrys. In this example, compounds are being selected based on the multiple criteria of the Lipinski rules.

example of a multiobjective optimization program. There is no consensus at present as to which tools and algorithms are the best for drug design. The following paragraphs give a discussion of some of the available options.

A fairly trivial way of representing multiple criteria is to have a page with a yes/no value noting whether each criterion is met. When this has been done, the results may be displayed, sorted by the total number of criteria met by each compound.

Another mechanism for making multiobjective criteria easily interpretable is to construct a single number that is a combination of multiple metrics. For example, Actelion uses a Drug Score value computed from the drug-likeness, ClogP, log *S*, molecular weight, and toxicity predictions. This is an improvement over the "yes/no" method described in the previous paragraph, because it gives a more quantitative measure, while continuing to have the simplicity of a single number on which to sort or select data. In some variations on this technique, the multiple criteria will be weighted to reflect the relative importance of each.

A technique similar in principle to the one in the previous paragraph is to do a histogram binning of each criterion. For example, each criterion may be given a rank of 1–5 based on how well that property fits the desired profile. These ranks can then be added to give an overall numerical score. Some researchers prefer this type of analysis, because two compounds that are essentially the same, within the accuracy of the scoring prediction, will end up having the same overall score. This avoids the pitfall of having a metric that makes it appear as though one compound is better than the other based on an insignificant difference.

There are a number of algorithms for walking through the space and finding an optimal point. These are in some ways analogous to the algorithms used for geometry optimization, although they must work in a space of discrete values, not the continuous space of possible molecular geometries. Some of the algorithms in use include homotopy techniques, epsilon-constraint methods, goal programming, normal-boundary intersection, and multilevel programming.

One of the basic multiobjective optimization functions is Pareto optimization, named after Vilfredo Pareto, who described it as a model of the way in which economic concerns lead to the most efficient distribution of income. It is a discrete formulation analogous to a downhill geometry optimization based on molecular energy.

One of the more heavily explored algorithms is MOGA (Multiobjective Genetic Algorithm), which applies evolutionary programming techniques to the multiobjective optimization problem. Internally, these algorithms almost always generate a single numerical value that gives an indication of whether one compound is better or worse than another. MoSELECT is a MOGA-based program that allows researchers to select the best compounds from large libraries based on multiple criteria.

This chapter has given a brief glimpse into the evolving area of multiobjective optimization. It is likely that new tools will be available in the near future. Researchers are advised to investigate the options presently available, and give consideration to how those tools might benefit their drug design work.

## BIBLIOGRAPHY

Agrafiotis DK. Multiobjective optimization of combinatorial libraries. IBM J Res Dev 2001; 45: 545–566.

Baringhaus KH, Matter H. Efficient strategies for lead optimization by simultaneously addressing affinity, selectivity and pharmacokinetic parameters. In: Oprea TI, editor. Chemoinformatics in Drug Discovery. Weinheim; Wiley-VCH: 2000. p. 333–379.

Drug Score. Organic Chemistry Portal. Available at http://www.organic-chemistry.org/prog/peo/drugScore.html. Accessed 2007 Sep 7.

Gillet VJ, Willett P, Fleming PJ, Green DVS. Designing focused libraries using MoSELECT. J Mol Graph Model 2002; 20: 491–498.

Handl J, Kell DB, Knowles J. Multiobjective optimization in bioinformatics and computational biology. IEEE/ACM Trans Comput Biol Bioinform 2007; 4: 279–292.

Murata T, Ishibuchi H. MOGA: multi-objective genetic algorithms. In: Proceedings of IEEE International Conference on Evolutionary Computation; 1995 Nov 29–Dec 1; Perth, WA, Australia. Volume 1. IEEE; 1996. p 289.

Nicolotti O, Gillett VJ, Flemming PJ, Green DV. Multiobjective optimization in quantitative structure-activity relationships: deriving accurate and interpretable QSARs. J Med Chem 2002; 45: 5069–5080.

A large list of references to evolutionary multiobjective optimization is available at http://www.lania.mx/~ccoello/EMOO/EMOObib.html.

Additional references are contained on the accompanying CD.

# 21

# AUTOMATION OF TASKS

In an industrial setting, drug design does not consist of creating a single molecule, or even a hundred molecules. Small drug design projects will test thousands of compounds, and large ones may test hundreds of thousands and even millions *in silico*. Because of this large scale, the automation of computational tasks is an important part of the job. The following paragraphs discuss some automation mechanisms found in chemistry software packages.

## 21.1 BUILT-IN AUTOMATION CAPABILITIES

Some chemistry software packages are built specifically to handle large numbers of compounds. This might be done by using a spreadsheet or database interface. For example, adding a column with some computed molecular property to a spreadsheet results in getting values for all of the compounds at once. The advantage of these programs is that no extra work is required to automate tasks for large numbers of compounds. The disadvantage is that the program may require more computer memory than would be used by automating a program that works on one molecule at a time. Figure 21.1 shows an example of a spreadsheet-like interface.

Some chemistry software packages have a built-in automation mechanism. There are a number of ways in which this is done, as illustrated in the following

*Computational Drug Design*. By David C. Young

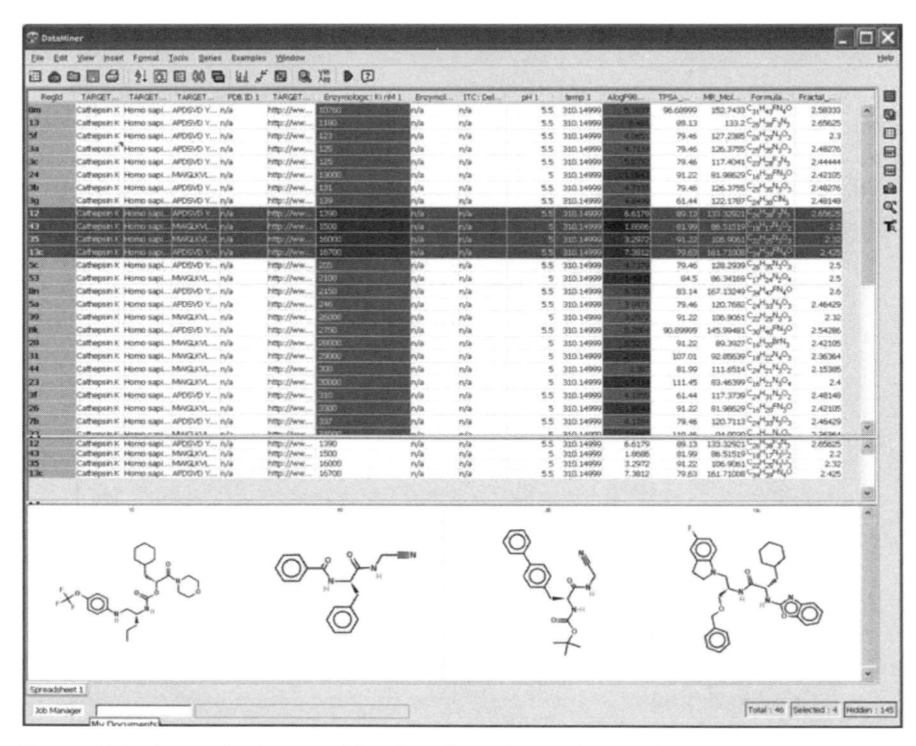

**Figure 21.1** Example of a spreadsheet interface. The application displayed here is Benchware DataMiner from Tripos, which has one of the most full-featured chemical spreadsheet formats.

examples. The Pipeline Pilot product from Scitegic is an icon-oriented programming interface in which icons representing various tasks are placed in a working area, and then connected together to show how data passes from one to the next. Several software packages use the Python programming language internally, or Jython (a Java implementation of Python that is used by the DataMiner program). The advantage of these systems is that they are very customizable, to the point that it is possible to create new functionality that is not part of the software package, in addition to automating the tasks that it normally performs. The disadvantage is that there can be a steep learning curve before a researcher becomes comfortable at writing code. A number of chemistry software companies, including Accelrys and Tripos, now offer Scitegic modules to access some of their product functionality inside Pipeline Pilot.

An extreme example of a package with built-in automation capacity is the MOE (molecular operating environment) drug design package from Chemical Computing Group. MOE ships with complete source code written

```
//     Bond atoms

       local numberOfAtoms = nAtoms Atoms[];

       for i=1,numberOfAtoms -1 loop
              aList = Atoms[];
              position = aPos aList;
              local position2 = aPos aList(i);
              local dist = add sqr (position - position2);
              local bound = aList | dist < 4.0 and aList <> aList(i);
              Bond [[aList(i)],[bound]];
       endloop
```

**Figure 21.2** A small section of code in Scientific Vector Language (SVL). Because this is a vector language, many variables are arrays, making it possible to easily work with all atoms in a molecule or other collections of similar data.

in Chemical Computing Group's Scientific Vector Language (SVL). SVL has been given a wide range of functions for manipulating chemical structure data, as well as numeric data. It is compiled on the fly to byte codes, as is Java. The license checkout is on the byte code interpreter, so a license must be purchased to run anything in SVL, whether created by the manufacturer or by someone else. This gives the manufacturer the freedom to distribute the full source code, and gives the user a platform that is both a fully functional drug design platform and an application development framework. Figure 21.2 shows an example of SVL code.

Another dedicated programming language is ChemBasic from Advanced Chemistry Development (ACD). ACD makes a suite of integrated cheminformatics tools. Their ChemBasic language has features of the BASIC computer language, and features reminiscent of Microsoft Visual Basic for Applications. It can be used to automate tasks within the ACD tools and to extend the functionality of those tools.

## 21.2 AUTOMATION USING EXTERNAL UTILITIES

Some chemistry software packages are built so that they can utilize the scripting capabilities in the operating system. For example, the UNITY tools from Tripos are a set of nongraphical executables that can be run from the UNITY or Linux command line. With this format, many tasks can be automated simply by putting a sequence of commands, "if" statements, loops, etc. into a Unix shell script. The advantage of this format is that it is flexible, automatable, and easy to integrate with other software packages that use a similar structure. The disadvantage is that scripts can easily become tied to a certain operating system, file naming convention, directory tree, etc.

The automation mechanisms mentioned thus far automate tasks from below, meaning that a program with a graphical user interface (GUI) is being automated by calling the underlying application programming interface (API). The alternative is to automate a GUI application from above, by automating the operating of the GUI. There are several mechanisms for doing this. A straightforward approach is to record a series of mouse clicks at specified points on the screen. This is not as ideal as it sounds, as the automation can break if run on a system with a different screen resolution, or when a new version of the software comes out with a slightly different screen format. Some GUI systems can be controlled through object-based drivers, which look for an interface to specify based on the type of widget (check box, pull-down menu, etc.), but may fail on custom widgets. A couple of very progressive, GUI programs have a "keyword testing" paradigm to externally drive the GUI in an unambiguous manner. None of these mechanisms is ideal, so GUI automation should be left as an option of last resort for situations when no other solution seems to be viable.

Automation is not the primary concern when selecting drug design tools. The ability to do the research to design a marketable drug will always be the first priority. However, part of that effort is the ability to do the volume of work needed for the project to stay on schedule. Since computational experiments are less expensive than laboratory synthesis and testing, organizations carry out virtual screenings on far more compounds than are actually synthesized and tested. Therefore, many computational drug design departments find that automation is the only way to meet the demands being made of them.

## BIBLIOGRAPHY

The best source of information on automating tasks with any particular software package is the documentation for that software. The following is a review of a utility for driving the graphical interface:

Open Testware Reviews: GUI Test Driver Survey. Tejas Software Consulting; 2003. Available at http://tejasconsulting.com/open-testware/feature/gui-test-driver-survey.html. Accessed 2007 Sep 21.

# PART III

# RELATED TOPICS

# 22

# BIOINFORMATICS

This chapter begins Part III of this book. This part provides simplified introductions to topics that are on the fringe of computational drug design. As such, these are things that a computational chemist may not actually do in an industrial setting, or may do only in rare instances. None the less, computational chemists should have a general understanding of these topics. The following discussion of bioinformatics and the discussions in the following chapters will be more general, high level discussions than are found in the rest of this book. The next step in delving into these subjects is provided by reading the works referenced at the end of the chapters; the step after that comes from reading the more advanced sources referenced in the supplemental references file on the accompanying CD.

It has been estimated that the amount of genomic data available doubles yearly. Each year, the journal *Nucleic Acids Research* publishes an issue devoted to genomic databases. In the past, this journal has presented a comprehensive listing of publicly available genomic databases, but not commercial or proprietary sources of genomic data. This provides a good starting point for finding out what genomic data is available.

All of the currently known drugs are based on fewer than 500 molecular targets. It is estimated that there may be anywhere from 1000 to 10,000 viable drug targets in the human genome. The current thinking in the pharmaceutical industry is that the process of identifying these targets for future drug

*Computational Drug Design*. By David C. Young

development will start with the human genome data, move on to identification of protein action (see Chapter 25), and then fit those proteins into an understanding of the body's interconnected metabolic pathways (the functional genomics movement). On an even more fundamental level, the emerging field of epigenetics examines how liver cells divide to create more liver cells, and how stem cells differentiate into other types of cells, even though all of these cells contain the exact same DNA sequences.

Another new research direction will be the analysis of genetic differences specific to various ethnic groups, and eventually individual patients. There are efforts underway to eventually make genome sequencing so cheap and easy that the genome of every individual patient can be sequenced. In time, this might give indications of predisposition to various diseases and hypersensitivity to specific drugs, and might possibly be used for more nefarious purposes, such as denying health insurance coverage. A host of bioinformatics tools are expected to be utilized in handling the large amounts of genomic data that must be processed in the course of these efforts.

In the rest of this chapter, the discussion will focus only on tools that are currently available. At present, there are a number of software tools for performing common tasks such as searching databases, aligning sequences, and creating phylogenetic trees. These low level tools can be utilized appropriately to perform a number of high level tasks, such as:

- determining the function of a protein;
- creating a 3D homology model of a protein;
- selecting the appropriate sequence to be targeted by an antisense drug;
- selecting an ideal animal model for drug efficacy testing;
- identifying proteins that should be assayed to check for risk of potential drug side effects;
- finding correlations between adverse effects of drugs and usage by selected ethnic groups;
- learning more about the metabolic pathways, evolution, genetics, and epigenetics relevant to a disease.

The most heavily performed function in bioinformatics is a search for other proteins that are similar to a query sequence. Most often, a newly identified protein is a member of a known family of proteins. Thus, finding proteins with known functions that are very similar to an unknown pretty reliably indicates the function of the unknown protein.

In the world of bioinformatics, there are precise measures of how similar one protein sequence is to another. The percent identity tells what percentage of the residues are exactly identical, ignoring gaps. In some of the older literature, there are references to a "percent similarity" or "percent homology" that count similar residues as matches: for example, valine and leucine, both of which are hydrophobic and similar in size. Authors taking a stricter interpretation will consider percent identity to be a factual measure of how similar two proteins are, and will refer to homology as a hypothesis that the two proteins share an evolutionary ancestor. As a practical matter, researchers will first consider percent identity, and then, if two matches have essentially the same percent identity, consider the one with more matches of similar residues (e.g., both hydrophobic) to be the better match. A sequence alignment can be performed to display the exact regions of the protein sequences that are identical, similar, or dissimilar, as shown in Fig. 22.1.

The process of homology model building is a topic that straddles the boundaries between computational chemistry and bioinformatics. Homology models

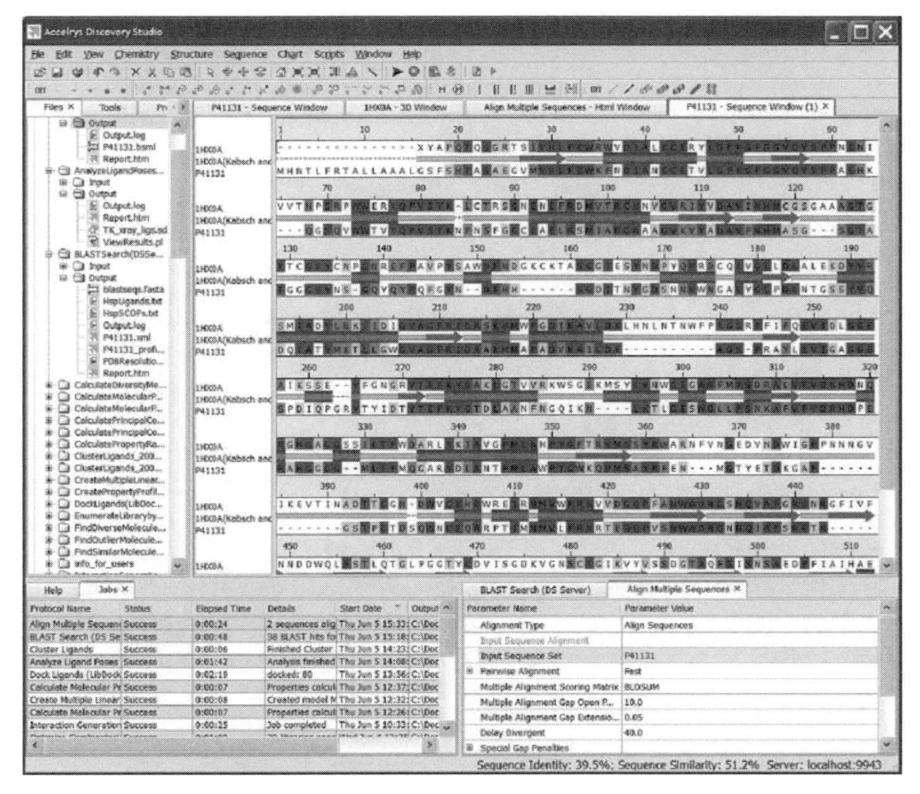


**Figure 22.1** A sequence alignment display from the Discovery Studio software from Accelrys. Identical residues are colored dark blue and similar residues light blue. The orange bars and blue arrows indicate regions where the secondary structure is predicted to be an α-helix or β-sheet. A color copy of this figure is contained on the accompanying CD.

may be created by computational chemists, biologists, or protein crystallographers, depending upon the corporate organization and skills of the individual researchers. For further discussion, see Chapter 9.

Antisense drugs are small RNA strands or RNA mimics that bind to RNA in order to prevent the translation process from generating a protein. The great attraction of antisense drug design is that knowledge of the DNA sequence gives enough information to immediately provide the structure of an antisense drug to downregulate the production of that protein. There are bioinformatics software programs that suggest the best antisense sequence to use to downregulate a particular protein. The major drawback of this design technique is that RNA strands make poor drugs owing to bioavailability concerns, and efforts to define bioavailable mimetics have been only minimally successful. Because of this, only one antisense drug (fomivirsen; trade name Vitravene, from Isis Pharmaceuticals) is currently approved in the United States for prescription use. In some cases, antisense compounds given by injection can be an alternative to creating a knockout mouse.

When choosing an animal model, the most important criterion is to find a species that is susceptible to the disease to be treated. Provided that several species meet this criterion, the next concern is to select one that will respond to the drug in the same way that a human will respond. Similarity metrics and phylogenetic trees (also called guide trees) can be used to choose which species is the most closely related in terms of only the target protein, as shown in Fig. 22.2. Cost is also a factor, as it is much less expensive to work with mice than larger primates, horses, etc.

One reason that drugs can exhibit unwanted side effects is that they might inhibit targets other than the intended protein. Many proteins are members of families of proteins with similar structures and functions. Thus, similarity metrics are used to identify the proteins that are most similar to the target protein. These proteins are often included in biochemical assays, even early in the drug design process, as an indication of a risk for unwanted side effects.

A number of drugs are known to affect patients of selected ethnic groups differently. For example, mild painkillers such as codeine have no effect on individuals of Arabian decent owing to enhanced detoxification metabolic pathways prevalent in this ethnic group. In the past, these drug efficacy profiles

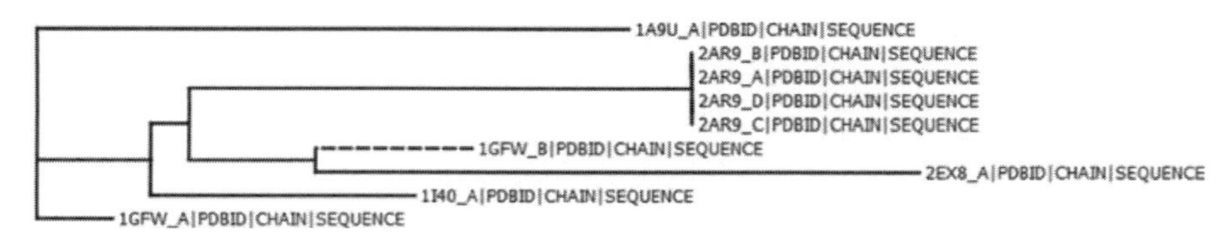


**Figure 22.2** A phylogenetic tree can be used to display the relative similarity of different proteins, or of the same protein from different species.

were determined by trial and error. In 2007, the first complete sequencing of a Han Chinese genome was completed. Other ethnic genomes are expected to become available in the future. It is expected that a standard part of the drug design effort in the near future will be to examine ethnic variation in the drug target, and targets identified as potential sources of unwanted side effects. This could allow drug designers to avoid ethnically related drug interactions, or perhaps develop drugs targeted to specific ethnic groups.

A drug target protein or RNA strand is not an isolated entity. It plays a role in a metabolic pathway, and often in multiple pathways. A given protein may also play slightly different roles at various stages in the growth, aging, and pregnancy of a patient. There are also evolutionary reasons that a protein came to be the way it is. Any understanding of any of these issues provides additional information that may be valuable to a drug designer. Gaining such understanding is a nontrivial task, and is at present at the cutting edge of the fields of biology and genetics.

On a more practical note, bioinformatics users will have to select appropriate software tools for their work. Many of these tools are currently available free in the public domain, and many can be accessed via free websites. Websites provide a simplified graphical front end to bioinformatics software. However, these programs often have additional functionality that is not accessible via web front ends. In order to utilize that additional functionality, the researcher will have to become familiar with work from a command line, often within the context of a Linux operating system, shell scripts, or a batch queuing system. Some corporations prohibit the use of public websites for research purposes, because of the risk of compromising the security of their proprietary data. The other advantage to running bioinformatics in house is that the researcher can control what version of the programs is being used, and thus be able to go back and reproduce results when desired. For this reason, many researchers permanently store all of their input settings, which is most conveniently done if they are running the software such that all of the inputs come from text files.

Bioinformatics is an important and growing field. It is already integrated into the industrial drug design process, and is likely to take on additional roles in the future. As such, a drug design researcher would not be wasting their time gaining some reasonable level of knowledge beyond that given in this chapter.

## BIBLIOGRAPHY

### Introductory Books

Barnes MR. Bioinformatics for Geneticists: A Bioinformatics Primer for the Analysis of Genetic Data. Hoboken, NJ: Wiley; 2007.

Baxevanis AD, Ouellette BFF, editors. Bioinformatics: A Practical Guide to the Analysis of Genes and Proteins. New York: Wiley; 2005.

Claverie JM, Notredame C. Bioinformatics for Dummies. Hoboken, NJ: Wiley; 2007.

Xiong J. Essential Bioinformatics. Cambridge: Cambridge University Press; 2006.

Crooke ST, editor. Antisense Drug Technology: Principles, Strategies, and Applications. Boca Raton: CRC Press; 2001.

Gad SC, editor. Handbook of Pharmaceutical Biotechnology. Hoboken, NJ: Wiley; 2007.

Krane DE, Raymer ML. Fundamental Concepts of Bioinformatics. San Francisco: Benjamin Cummings; 2002.

Lesk AM. Introduction to Bioinformatics. Oxford: Oxford University Press; 2005.

Zvelebil M, Baum J. Understanding Bioinformatics. Oxford: Garland Science; 2007.

## Review Article

Esposito EX, Tobi D, Madura JD. Comparative protein modeling. Rev Comput Chem 2006; 22: 57–167.

## Discussion of the Number of Potential Drug Targets

Drews J. Drug discovery: a historical perspective. Science 2000; 287: 1960–1964.

Additional references are contained on the accompanying CD.

# 23

# SIMULATIONS AT THE CELLULAR AND ORGAN LEVEL

As computer hardware becomes more powerful and the functionality of available software packages becomes greater, there is a trend towards running more complex simulations. This trend has moved computational chemistry from simulations of atoms to simulations of increasingly larger molecules and proteins. Today, there is even software that does simulations, at an appropriate level of theory, on the entire human body.

It is expected that this trend will continue. This expectation is based on the fact that increasingly complex and large simulations hold a promise of new improvements to benefit drug design work. For example, at present, oral bioavailability is typically predicted with a QSAR model of passive intestinal absorption. However, a more complex simulation could identify molecules that will be transported via an active transport mechanism and make a more accurate prediction of their bioavailability to specific organs.

## 23.1 CELLULAR SIMULATIONS

Molecular mechanics and molecular dynamics calculations on proteins are now commonplace (although not necessarily trivial to perform). Many molecular properties are primarily computed with QSAR techniques. In rare cases, simulations of proteins in membranes, or of complexes of multiple proteins, have been performed. Only in the rarest cases are there interconnected simulation systems that utilize multiple types of simulations or calculations in order to give an overall effect of an entire biological system, such as a metabolic pathway or an organelle.

As researchers move to simulations of larger and larger systems, there is a scale issue to be negotiated. Each type of simulation is practical only up to a certain size of system. For example, some researchers have published simulations of a section of cell membrane with a membrane protein. At present, such simulations are beyond the capability of quantum mechanical approaches, which simulate each electron and atomic nucleus. These simulations are approaching the limits of what can be done with molecular mechanics and molecular dynamics, which simulate each atom and bond. In a few rare cases, researchers have found it a viable option to simulate a membrane as a sheet of continuous material, very similar to how a structural engineer would simulate a sheet of rubber. In between these last two approaches are mesoscale calculations, which may simulate a lipid as a string of beads or some other geometric shape such as a cylinder. On an even larger scale, QSAR methods can be used to create algebraic models of entire organs, such as a model to predict how rapidly the liver will eliminate a drug from the bloodstream.

Most atom level membrane simulations are performed using standard molecular mechanics or Monte Carlo software packages. However, one alternative is illustrated by the COSMOmic program from COSMOlogic. This is a program designed specifically to model micelles and biomembranes, either as spherical micelles or as planar layers of inhomogeneous liquids. This is sometimes done using a spherical representation with an appropriate potential, as shown in Fig. 23.1. This can be used to compute membrane

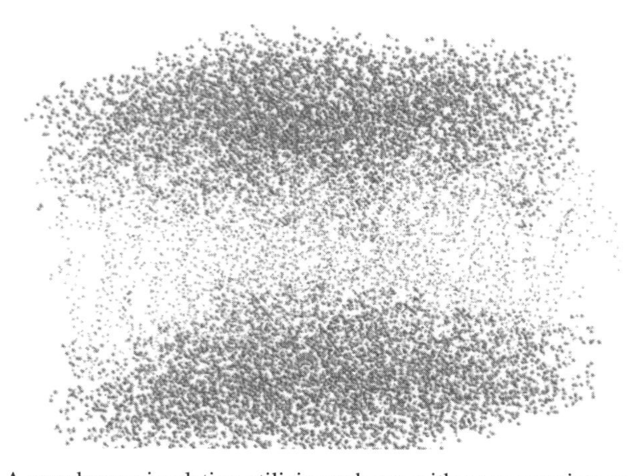

**Figure 23.1** A membrane simulation utilizing spheres with an appropriate potential function.

partition coefficients of solutes in micelles or biomembranes, and probability distribution functions.

The boundaries defining which type of model is best for a given simulation, or even possible, slowly expand as computer software and computing hardware undergo a slow evolution. Regardless of this evolution, it seems likely that no single type of model will be ideal for all simulations any time in the near future. Therefore, various types of models and simulations that combine multiple models are most likely here to stay. This raises a number of issues of scientific concern. The first is the validation of the accuracy and applicability of each technique. The second is the creation of mechanisms to interconnect those models, and, more importantly, validate that the interconnected simulation gives reasonable results.

Within the field of computational cell biology, nearly all of the simulations are dynamical in nature. These simulations mimic biochemical pathways, neurons firing, transporters, pumps, intercellular communication, circadian rhythms, and other processes. These can be biochemical pathway simulations, whole cell models, or multiple cell models. Biochemical pathway models usually consist of a system of differential equations that are interconnected through having the species created by one being utilized by another. Whole cell models generally incorporate the spatial arrangement of features within the cell and account for diffusion of molecular species from one enzyme to the next.

A number of programs have been created for simulation of biochemical metabolic pathways, based on numerical integration of rate equations. These include GEPASI, KINSIM, MIST, METAMODEL, GEISHA SYSTEM, and SCAMP. The step beyond this is the simulation of the concentrations of all of the species (both protein and small molecule) in a cell with multiple metabolic pathways or an entire genome. E-Cell System is a rule-based simulation system written in C++. E-Cell uses reaction rate equations to model enzyme actions. It can be used to watch the concentrations of species change over time as it progresses with discrete time steps that model all enzymes working in parallel.

StochSim, MCell, and Virtual Cell are based on 3D spatial arrangements and stochastic simulations. In these programs, a random number generator is used to determine which enzyme undergoes a reaction at each iteration. This has the advantage of searching a very complex space of possible states in an efficient manner. Note that the Virtual Cell is the software package developed by the National Resource for Cell Analysis and Modeling (NRCAM) and hosted at http://www.vcell.org/. This is not to be confused with several other "virtual cell" websites, which tend to be online tutorials primarily aimed at the education of high school students, rather than simulations.

## 23.2 ORGAN SIMULATIONS

At a higher level, there are simulations of entire organs. These simulations tend to fall into two categories:

- physical simulations, such as the flow of blood or air through the organ
- metabolic simulations designed to mimic how the organ is responsible for allowing new nutrients into the blood plasma, eliminating wastes from the body, etc.

There have been simulations of air flow through the lungs and nasal cavities that are done purely as computational fluid dynamics (CFD) simulations. Similarly, the flow of blood through the heart, arteries, and veins can be simulated as the flow of a continuous fluid. In some cases, the organs themselves are held rigid; in other cases, movement is simulated, such as the beating of a heart. One could foresee that someday such flow calculations may be included in a prediction of how rapidly a drug is transported to its target in the body.

On a larger scale, the GastroPlus software (see Fig. 19.2) from Simulations Plus uses a model of the entire digestive tract in order to make ADMET predictions. Perhaps the most ambitious modeling project thus far is the Digital Human project. The Digital Human is an open source software consortium building a framework for modeling a human on multiple scales ranging from DNA up through cells and organs to gross anatomy.

The larger scale simulations introduced in this chapter are a small, but growing, segment of the modeling work in the pharmaceutical research and development field. It is likely that new simulation tools will become available and more widely used. Most likely, these tools will be based on the same types of computations discussed here. The majority of these tools are still in the early stages of development. None the less, drug design researchers should have some familiarity with the large scale modeling tools that are relevant to their work, so that they will have some indication of when it becomes advantageous to start utilizing them.

## BIBLIOGRAPHY

### Membrane Simulation Reviews

Biggin PC, Sansom MS. Interactions of alpha-helices with lipid bilayers: A review of simulation studies. Biophys Chem 1999; 76: 161–183.

Bond P, Sansom M. The simulation approach to bacterial outer membrane proteins (Review). Mol Membr Biol 2004; 21: 151–161.

Burrage K, Hancock J, Leier A, Nicolau DV Jr. Modeling and simulation techniques for membrane biology. Brief Bioinform 2007; 8: 234–244.

Chiu SW, Clark M, Balaji V, Subramaniam S, Scott HL, Jakobsson E. Simulation of a fluid phase lipid bilayer membrane: Incorporation of the surface tension into system boundary conditions. Mol Eng 1995; 5: 45–53.

## Cell Simulation Books and Reviews

Alt W, Chaplain M, Griebel M, Lenz J, editors. Polymer and Cell Dynamics: Multiscale Modeling and Numerical Simulations. Basel: Birkhäuser; 2003.

Arita M, Hagiya M, Shiratori T. GEISHA SYSTEM: An environment for simulating protein interaction. Genome Inform 1994; 5: 80–89.

Cremer T, Kreth G, Koester H, Fink RH, Heintzmann R, Cremer M, Solovei I, Zink D, Cremer C. Chromosome territories, interchromatin domain compartment, and nuclear matrix: An integrated view of the functional nuclear architecture. Crit Rev Eukaryot Gene Expr 2000; 10: 179–212.

Fall C, Marland E, Wagner J, Tyson J, editors. Computational Cell Biology. New York: Springer; 2002.

Loew LM, Schaff JC. The Virtual Cell: A software environment for computational cell biology. Trends Biotechnol 2001; 19: 401–406.

Pozrikidis C, editor. Modeling and Simulation of Capsules and Biological Cells. Boca Raton, FL: CRC/Chapman & Hall; 2003.

Slepchenko BM, Schaff JC, Carson JH, Loew LM. Computational cell biology: Spatiotemporal simulation of cellular events. Annu Rev Biophys Biomol Struct 2002; 31: 423–441.

## Organ Simulation Reviews

Bassingthwaighte JB. Strategies for the physiome project. Ann Biomed Eng 2000; 28: 1043–1058.

Dao N, McCormick PJ, Dewey CF Jr. The human physiome as an information environment. Ann Biomed Eng 2000; 28: 1032–1042.

Digital Human. Available at http://www.fas.org/dh/. Accessed 2007 Dec 3.

Noble D. Modeling the heart—from genes to cells to the whole organ. Science 2002; 295: 1678–1682.

Additional references are contained on the accompanying CD.

# 24

# SYNTHESIS ROUTE PREDICTION

Organic and medicinal chemists can spend a significant amount of time developing synthesis routes, particularly for complex compounds such as natural products. There is a class of computer programs designed to automatically suggest a synthesis route. These programs are often referred to as CAOS (computer aided organic synthesis) programs or sometimes as SDS (synthesis design systems) programs. None of them is the be all and end all of organic chemistry. However, they can suggest useful synthesis routes and save time on literature searching.

The first CAOS program, OCSS (Organic Chemistry Synthesis Simulator) was created by E. J. Corey and T. Wipke. Some of the other software packages that have been developed are described in Table 24.1. Many of the available programs were designed to test an algorithmic design by making them able to synthesize just a single type of compound, such as DNA, heterocyclic systems, or phosphorous compounds. A few programs have been designed to suggest a synthesis for any organic molecule. These programs use one of a few software architectures. Most work retrosynthetically to go backwards from the intended product to compounds that are commercially available. A few start with the most similar available starting compound and work forwards to the desired product.

Some programs use a database of known chemical reactions. The advantage of the database approach is that it suggests routes in which all of the steps are

*Computational Drug Design*. By David C. Young

**TABLE 24.1 CAOS Software Packages**

| Name | Description |
|---|---|
| ARChem | Retrosynthetic database, based on a rule extraction engine |
| CAESA | Both synthetic and retrosynthetic; causal network expert system |
| CAMEO | Forward synthesis indicates feasibility of individual steps |
| CASINO | Retrosynthetic from target to designated starting material; heuristics |
| CHIRON | Optimal functional groups; forward synthesis |
| EROS | Formal; heuristics limit combinatorial explosion; bond enthalpies; retrosynthetic |
| HOLOWin | Both retrosynthetic and forward synthesis alternatives; database |
| IGOR | Formal; treats reactions as isomerism |
| LHASA | Retrosynthetic; heuristic; knowledge base; interactive; alternatively, starting material can be specified |
| OCSS | Retrosynthetic; heuristic; knowledge base |
| OSET | Open source; retrosynthetic |
| SECS | Offshoot of LHASA; improved stereochemistry and double bond geometry |
| SESAM | Formal; retrosynthetic |
| SST | Formal; pattern recognition to choose starting material |
| SYNCHEM | Noninteractive; retrosynthetic; improved AI and heuristics |
| SYNCHEM2 | Multiprocessor; improved reaction description |
| SYNGEN | Formal; symbolic representation of bond formation/cleavage; shortest route synthesis |
| Syntree | Database; retrosynthetic |
| WODCA | Both retrosynthetic and forward synthesis from best precursors using database |

known reactions. The disadvantages are that the results are only as good as the database and that there is no possibility of suggesting a novel reaction. Such programs require some level of commitment on the part of the users or developers to constantly update the database. One solution to this is to use a web-based implementation such as that shown in Fig. 24.1 where the most recent reaction set is always available online.

Other CAOS programs utilize a set of heuristic rules about how to select critical bonds for synthesis, attachment points, etc. The advantage of a program that does not utilize a reaction database is that it can suggest novel, previously unknown, synthesis strategies. The disadvantage is that it may suggest something that is not feasible. The algorithms within these programs may be expert systems, heuristic rules, or other types of artificial intelligence algorithms.

A third architecture is represented by programs that combinatorially try many different options for selecting which sequence of bonds to form or break. This can find synthesis routes that were not within the suggestions included in the artificial intelligence program. The disadvantage is that it may generate a very large number of synthesis routes, many of which are not practical.

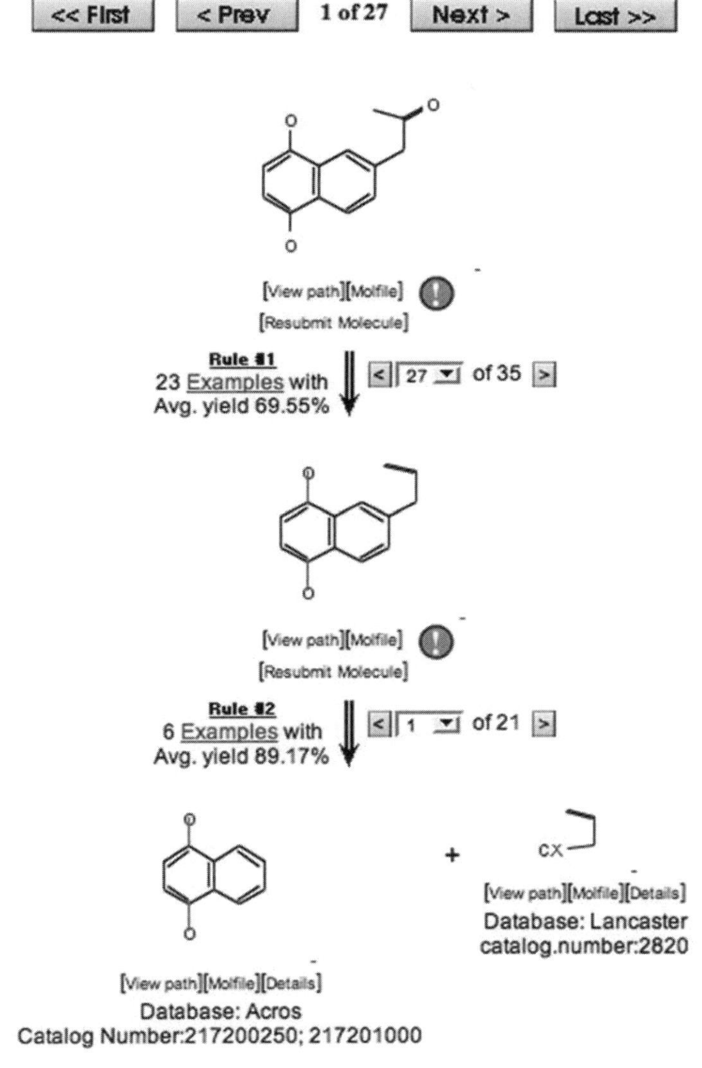


**Figure 24.1** Sample results from ARChem, a web-based synthesis route prediction program from SimBioSys.

The number of groups developing CAOS seems to have decreased in recent years. This author will leave it for others to speculate as to whether these efforts have hit a theoretical snag, or if it is simply a currently unpopular subject with grant funding agencies. In either case, the existing tools can sometimes be of value.

## BIBLIOGRAPHY

Barone R, Chanon M. Synthesis design. In: von Ragué Schleyer P et al., editors. Encyclopedia of Computational Chemistry. New York: Wiley; 1998. p. 2931–2948.

Todd MH. Computer-aided organic synthesis. Chem Soc Rev 2005; 34: 247–266.

Wipke WT, Howe WJ, editors. Computer-Assisted Organic Synthesis. Washington DC: ACS; 1977.

Young D. Computational Chemistry: A Practical Guide for Applying Techniques to Real World Problems. New York: Wiley; 2001.

Additional references are contained on the accompanying CD.

# 25

# PROTEOMICS

Proteomics is one of the newest fields discussed in this text. At present, there are a number of definitions of proteomics that are not completely in agreement. In simplest terms, proteomics is the study of an organism's complete complement of proteins. Many proteomics projects focus specifically on the human proteome. Proteomics is the next natural step beyond the Human Genome Project. The following are some of the areas of research that have sometimes been cited as goals of the proteomics movement:

- Identify what proteins there are in a given organism.
- Determine how each protein is produced.
- Determine how proteins interact.
- Understand which proteins are expressed where and why.

The question of how each protein is produced leads to the study of a complex set of reactions. There are about 35,000 open reading frames in the DNA, which lead to the creation of the order of 500,000 proteins. This explosion in the number of protein species is due to posttranslational modifications (PTMs). Some PTMs are due to cleavage of proteins by proteases or to combination for multiple protein strands into a single strand. Other PTMs involve changes in the structure of amino acids. Perhaps the most frequently observed PTMs are reactions that add new functional groups onto the protein strands. These

*Computational Drug Design*. By David C. Young

include reactions such as acylation, biotinylation, addition of heme groups, and addition of complex carbohydrate structures.

The question of how proteins interact opens up a range of issues. This vague question can encompass metabolic pathways, formation of functional protein units from multiple strands, and the processes by which proteins are discarded and metabolized.

The Human Proteome Organisation (HUPO) is an international scientific organization that promotes and sponsors proteome work. HUPO is currently sponsoring nine proteomics initiatives focusing on the liver, brain, heart, antibodies, mouse models, standards, plasma proteome, disease proteomes, and stem cells. HUPO is funded by governmental, academic, and industry partners.

Clearly, there are overlaps between the goals of the proteomics movement and those of structural genomics, functional genomics, and other initiatives. Indeed, the proteomics field is sometimes described as being subdivided into branches such as protein separation, protein identification, quantitative proteomics, protein sequence analysis, structural proteomics, interaction proteomics, protein modification, cellular proteomics, membrane proteomics, and experimental bioinformatics. Thus far, the primary benefit of the fledgling proteomics movement has been finding biomarkers to measure disease progression and drug efficacy. Over time, the field may define itself differently, but it seems certain that studying proteins will continue to be a widespread and important field of work in the future.

At present, it is unclear what shape the field of proteomics may take in the future. Regardless of how the field evolves, it is likely to find an important role in the pharmaceutical industry.

## BIBLIOGRAPHY

Anderson NL, Anderson NG. Proteome and proteomics: New technologies, new concepts, and new words. Electrophoresis 1998; 19: 1853–1861.

Blackstock WP, Weir MP. Proteomics: Quantitative and physical mapping of cellular proteins. Trends Biotechnol 1999; 17: 121–127.

Campbell AM, Heyer LJ. Discovering Genomics, Proteomics, and Bioinformatics. 2nd ed. San Francisco: Benjamin Cummings; 2006.

Hamacher M, Marcus K, Stühler K, van Hall A, Warscheid B, Meyer HE, Mannhold R, Kubinyi H, Folkers G, editors. Proteomics in Drug Research. Hoboken, NJ: Wiley; 2006.

Human Proteome Organisation website. Available at http://www.hupo.org. Accessed 2007 Dec 19.

Liebler DC. Introduction to Proteomics: Tools for the New Biology. Totowa, NJ: Humana Press; 2002.

Posttranslational Modification. Wikipedia. Available at http://en.wikipedia.org/wiki/Posttranslational_modification. Accessed 2007 Dec 18.

Proteomics. Wikipedia. Available at http://en.wikipedia.org/wiki/Proteomics. Accessed 2007 Dec 14.

Twyman RM. Principles of Proteomics. New York: BIOS Scientific Publishers; 2004.

Veenstra TD, Yates JR. Proteomics for Biological Discovery. Hoboken, NJ: Wiley-Liss; 2006.

Wang Y, Chiu JF, He QY. Proteomics in computer-aided drug design. Curr Comput Aided Drug Des 2005; 1: 43–52.

Additional references are contained on the accompanying CD.

# 26

# PRODRUG APPROACHES

If at all possible, drugs are designed to be orally bioavailable, so that they can be administered in tablet form. Unfortunately, this is not always possible. Creating a compound that will have sufficiently high efficacy usually dictates that most of the structure of the drug must fit within a very tight window of size and functional group properties. As such, there may be only one or two functional groups that can be altered slightly in order to adjust the physical properties to give an orally bioavailable drug. Thus, there are many situations where there is no possible way to make the drug both have the desired efficacy and be orally bioavailable.

When a drug cannot be made orally bioavailable, one alternative is to use a different treatment regime. If a single dose is needed, the drug can be administered by injection. A drug can be given intravenously, which is a viable approach if the drug will be administered for a short time and in a situation that requires the patient to be hospitalized. If the drug is to be given chronically, it may be possible to design a time-release device to be implanted in the patient. In many cases, injection and implants are not viable options, and thus a prodrug approach is the best solution to the problem.

A prodrug is a compound that can be administered (usually orally) but does not itself have any drug efficacy. However, after the compound enters the body, it is metabolized. One of the metabolites is then the active drug. Thus, a prodrug is actually a precursor to a drug. This can be an excellent solution to a bioavailability problem. In fact, some compounds that were discovered by trial and error (e.g., codeine) were later found to be prodrugs once their metabolism had been determined.

Usually, the prodrug is formed from the active drug covalently bound to another chemical moiety. That covalent bond is broken, usually enzymatically, once the prodrug has entered the bloodstream, target organ, or target cell. In some publications, the term "prodrug" is applied to a salt of the active drug.

The majority of prodrugs are formed by replacing an alcohol or carboxylic acid group with an ester linkage to some other group, resulting in the compound as a whole having better bioavailability characteristics. The ester can be hydrolyzed by stomach acid, but the desired behavior is generally for the more lipophilic prodrug to enter the bloodstream via passive intestinal absorption in the small intestine, and then being cleaved by an esterase closer to the drug target. Other prodrug activation methods include cleavage of amides,

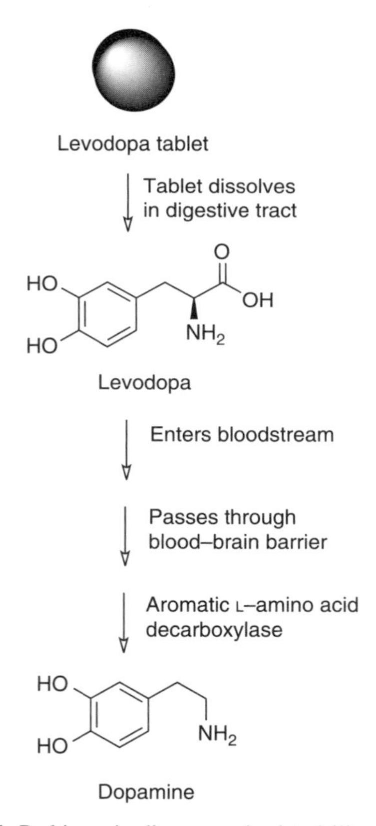


**Figure 26.1** Patients with Parkinson's disease and related illnesses have lowered levels of dopamine in the brain. Dopamine cannot be administered orally, because it does not cross the blood–brain barrier. Levodopa is a prodrug form of dopamine.

acyloxyalkyl esters, carbonate esters, phosphate esters, ethers, carbamates, imines, and enamines.

In some cases, a prodrug approach can produce some improvement in the efficacy and specificity of the drug. This happens when the enzyme that metabolizes the prodrug is primarily found in the organ or cells in which the drug target is located. For example, there are some cancer chemotherapy prodrugs that are primarily metabolized in the tumor cells, thus giving the desired efficacy with significantly fewer adverse side effects than non-prodrug administration of the compound.

There are a couple of other reasons for choosing a prodrug approach, in addition to the active form not being orally bioavailable. A prodrug alternative can be necessary when the active compound is chemically unstable in some biological environments (e.g., the stomach), and must therefore be created just before reaching the target. If the desired method of administration is to apply a "patch" to the patient's skin, then the best option is often to make a prodrug that will first be able to permeate the skin, and then be converted to the active form of the drug. A prodrug form may also be necessary to allow a drug to pass through the blood–brain barrier (Fig. 26.1).

In most cases, the functional group that is bonded to the active drug to form the prodrug is simply a group with the desired physicochemical properties. This group must be considered to make sure that it is something that will be

**TABLE 26.1 Examples of Prodrugs**

| | | |
|---|---|---|
| Adrafinil | Etofibrate | Prednisone |
| Alatrofloxacin | Famciclovir | Proglumetacin |
| Amifostine | Fosamprenavir | Proguanil |
| Avizafone | Fosfluconazole | Pyrazinamide |
| Azathioprine | Fosphenytoin | Quinapril |
| Bambuterol | Heroin | Rabeprazole |
| Benazepril | Hetacillin | Ramipril |
| Bezitramide | Indometacin farnesil | Rilmazafone |
| Capecitabine | Irinotecan | Ronifibrate |
| Carbamazepine | Isoniazid | Sodium phenylbutyrate |
| Carisoprodol | Levodopa | Spirapril |
| Clofibrate | Loxoprofen | Sulindac |
| Clofibride | Melevodopa | Taribavirin |
| Codeine | Mestranol | Temozolomide |
| Cyclophosphamide | Methyl aminolevulinate | Tenatoprazole |
| Dipivefrine | Nabumetone | Terfenadine |
| Dirithromycin | Nitazoxanide | Trandolapril |
| Dolasetron | Parecoxib | Valaciclovir |
| Enalapril | Pivampicillin | Valganciclovir |
| Etilevodopa | Potassium canrenoate | Ximelagatran |

harmlessly metabolized by the body. In some cases, the choice of precursor species can help with targeted drug delivery, such as attaching the drug to a monoclonal antibody or a structure that targets a membrane transport protein. This latter example is referred to as a "targeted prodrug" approach. A counter-intuitive example is that some antitumor drugs target tumor cells preferentially because the drugs are metabolized slowly in the tumor cells and more quickly in healthy cells.

Typically, drug design projects do not start out as prodrug design projects. Most often, they start out by creating a small molecule optimized to maximize activity in biochemical assays. The need for a prodrug formulation then becomes evident as the project moves on to cell-based testing or animal testing. There will often be both cell-based assay data and computational predictions of bioavailability to indicate the need for a prodrug formulation.

Prodrugs have been proven as successfully marketed therapeutics. Table 26.1 shows some examples of existing prodrugs. In light of the number of drug design projects that have successfully developed prodrugs, a prodrug option should be kept in every drug designer's bag of tricks.

## BIBLIOGRAPHY

Bhosle D, Bharambe S, Gairola N, Dhaneshwar SS. Mutual prodrug concept: Fundamentals and applications. Indian J Pharm Sci 2006; 68: 286–294.

de Albuquerque Silva AT, Chung MC, Castro LF, Carvalho Guido RV, Ferreira EI. Advances in prodrug design. Mini Rev Med Chem 2005; 5: 893–914.

Han HK, Amidon GL. Targeted prodrug design to optimize drug delivery. AAPS Pharm Sci 2000; 2: 1–11.

Majumdar S, Duvvuri S, Mitra AK. Membrane transporter/receptor-targeted prodrug design: strategies for human and veterinary drug development. Adv Drug Deliv Rev 2004; 56: 1437–452.

Melton RG, Knox RJ, editors. Enzyme-Prodrug Strategies for Cancer Therapy. New York: Springer; 1999.

Rooseboom M, Commandeur JNM, Vermeulen NPE. Enzyme-catalyzed activation of anticancer prodrugs. Pharmacol Rev 2004; 56: 53–102.

Rosen MR, editor. Delivery System Handbook for Personal Care and Cosmetic Products: Technology, Applications and Formulations. Norwich: William Andrew; 2005.

Silverman RB. Organic Chemistry of Drug Design and Drug Action. San Diego: Academic Press; 2004.

Stella VJ, Borchardt RT, Hageman MJ, Oliyai R, Maag H, Tilley J, editors. Prodrugs: Challenges and Rewards. New York: Springer; 2007.

Stella VJ, Nti-Addae KW. Prodrug strategies to overcome poor water solubility. Adv Drug Deliv Rev 2007; 59: 677–694.

Testa B, Mayer J. Prodrug design. In: Swarbrick J, Boylan JC, editors. Encyclopedia of Pharmaceutical Technology. New York: Marcel Dekker; 2002. p. 2304–2311.

Additional references are contained on the accompanying CD.

# 27

# FUTURE DEVELOPMENTS IN DRUG DESIGN

Any text on a scientific field is doomed to be obsolete sooner than most other books. Some of this obsolescence is because the topics discussed in the text change over time, and some is because new topics become important and must be added to future editions. This chapter suggests some of the areas of the field that currently appear to be looming in the future. This is being done to suggest that researchers keep an eye on developments in these areas. They are also being mentioned to give a brief introduction as to how these new topics fit with the other information in this book.

## 27.1 INDIVIDUAL PATIENT GENOME SEQUENCING

Since the completion of the Human Genome Project, DNA sequencing has continued to become faster and cheaper. Several companies have stated that they hope to make it possible to have complete genomes for individual patients, eventually at a cost of around $1000 per patient. This would probably first be utilized in clinical trials, to see if there are correlations between patients having adverse side effects and particular genetic traits.

Once individual genome sequencing becomes part of clinical trials, it is likely to become widely used for the population as a whole. The primary driving factor may be liability. If a doctor has a genome for a patient, they can

*Computational Drug Design*. By David C. Young

analyze for predisposition to many diseases, as well as possible adverse drug reactions. Consumer rights groups have already been lobbying for legislation making it illegal for insurance companies to deny health insurance coverage based on patient genome testing.

***27.1.1.1 Author's Side Note*** When writing began on this text, individual patient sequencing seemed to be a pipe dream that could be too expensive or take decades to accomplish. Two years later, it now appears that first clinical trials utilizing patient genome sequencing could be performed (if one can believe the marketing claims) nearly before the ink is dry on this book.

## 27.2 ANALYSIS OF THE ENTIRE PROTEOME

One concern in drug design is the possibility that a drug may have unintended, severe side effects. One way to analyze for this potential is to see if the drug binds to active sites in proteins other than the intended target. In theory, this can be done by automating the task of docking the drug compound with every active site in every protein in the body. Only a few companies are currently exploring the feasibility of this type of analysis. However, it would not be surprising if it became widespread in the future. The driving force for this movement would be the massive cost savings in avoiding a failed human clinical trial.

The next logical step is to map out where each of these proteins fits in the metabolic pathways. This, in turn, will give information about which proteins are expressed in which tissues, during pregnancy, etc. Or perhaps it will be the other way around, and understanding expression will give insight into metabolic pathways. For more information, see Chapter 25.

## 27.3 DRUGS CUSTOMIZED FOR ETHNIC GROUP OR INDIVIDUAL PATIENT

In the fall of 2007, the genome was completed for an individual of Han Chinese descent. This and similar projects can open the way to the development of drugs specifically targeted to the nuances in biology associated with a specific ethnic group. There are a number of diseases that have much higher incidences in selected ethnic groups. There are also a number of identified drugs that cause adverse reactions in members of specific ethnic groups. Therefore, development of drugs tailored to specific ethnic groups could be a profitable business strategy. It is likely that cost concerns would limit ethnic group-specific drug design projects to the more populous ethnic groups.

Once it is possible to sequence the genome for each patient, it becomes possible to individually design drugs for each patient. This is in the realm of possibility, but not necessarily feasibility. One issue is the cost. It would be necessary to have some highly automated way of designing and synthesizing custom drugs. At present, antisense compounds seem to be the best hope for this type of automated drug production.

An even larger issue is that of regulatory approval of the process by, for example in the United States, the FDA. Under current regulations, a drug tailored to a specific patient would still have to go through clinical trials, which is, in itself, impossible, because multiple patients are needed to conduct trials. Thus, in order to make drugs tailored to individual patients, it would be necessary to put a whole different type of approval process in place.

## 27.4 GENETIC MANIPULATION

Initial attempts at gene replacement treatment for genetic disorders have met with mixed success. The fact that there have been some successes suggests that this will be technically feasible on a widespread basis, although how far in the future remains to be seen. However, the potential treatments will meet with a myriad of personal, social, ethical, and religious issues. This is not a simple yes or no situation. It is a situation where individuals, society, and ultimately government will have to choose at what point to draw the line. Consider the following scenarios, any of which is likely to be one where gene replacement is the only possible treatment:

- Your baby is diagnosed with Tay–Sachs disease, which is a certain death sentence before their 5th birthday.
- Your baby is diagnosed with Down's syndrome. While not life-threatening, this disease will affect their physical, mental, and social development for their entire life.
- Your baby is diagnosed with dwarfism. As well as the social implications of being very short, they are destined to have many health problems and ultimately die younger than average owing to this condition.
- Nearly all of your family members have died at a young age from heart disease. There is a gene replacement that can make your body impervious to cholesterol and greatly cut down on heart disease. (Such a genetic mutation does exist. Everyone who has it is a descendent of one boy born in Europe in the Middle Ages.) Do you have yourself or your child treated?

- Nearly all of your family members are under 4′ 6″ tall (1.37 m). Do you allow your unborn fetus to undergo gene therapy to obtain an average height?
- There are gene replacements that will allow your children to have high intelligence and astounding physical aptitude. Do you alter their genes to give them an improved chance at succeeding in life? What if 90% of the other students in their school classes will have had the same enhancements? In this situation, not having your child's DNA altered would likely relegate them to the special education class.

All of these situations may one day become realities. Some are probably many years away. However, they deserve consideration now, because right now special interest groups are sponsoring legislation in attempts to either force choices on the population or remove those options from them.

## 27.5 CLONING

Cloning has received a large amount of attention in recent years, ever since the first cloning of a mammal, a sheep named Dolly. Advocates of cloning technology suggest that it could be possible to grow replacement organs that would not be rejected by the patient's body, because they are genetically the patient's own tissues. Opponents of cloning are afraid that living, cloned people will be killed just to use them as organ donors for wealthy patients. As the political and moral debates carry on, some of the most cutting edge research may be getting done in developing countries, which traditionally lag in human rights legislation. Whether this is a bad thing because of ethical concerns, or because it means that the largest countries are no longer the world's technology leaders, is up to the reader's point of view.

The cloning that attracts media attention is "reproductive cloning," in which a copy of an entire living organism is made. This is the only type of cloning that most of the nonscientific community have heard of, but there are two other types of cloning in use in the scientific community. "DNA cloning" is the process of copying a DNA sequence so that it can ultimately be inserted in a different cell line. DNA cloning is also called "molecular cloning," or "gene cloning," or "recombinant DNA technology." "Therapeutic cloning" is a means for generating stem cells from a fertilized egg. After 5 days of growth, the cells are extracted from the egg, which prevents any further growth. Therapeutic cloning is also called "embryo cloning."

Reproductive cloning of animals has been somewhat successful. Some species, such as horses, dogs, chickens, and primates, have not been cloned successfully and seem resistant to the same processes used to clone Dolly the sheep. Other species, including mice, pigs, goats, cows, cats, sheep, and several endangered animals, have been cloned with similar techniques. Even at its most successful, reproductive cloning is a very delicate process (Dolly was the single success out of 176 attempts). The cloned animals tend to have higher than average birth weight and some unexplained health problems. There have been several reports of cloned animals having shorter lifespans than their naturally produced siblings. Humans have not been cloned in this way, and there is evolving legislation and scientific ethical guidelines against doing this; however, there have been reports of failed human reproductive cloning experiments conducted in developing countries.

The ability to clone organs could have a massive impact. Currently, transplant patients must take drugs to suppress their immune system so that the organs will not be rejected. Subsequently, 90% of transplant patients get skin cancer due to their weakened immune system, compared with the 20% of the overall population that get skin cancer sometime in their life. At present, however, there are still significant technical barriers to overcome before organ cloning can become a reality.

## 27.6 STEM CELLS

Stem cells are undifferentiated cells. In a developing fetus, all cells are the same, until some little understood chemical trigger causes each cell to proceed down the correct developmental path to become a bone cell, muscle cell, nerve cell, etc. The hope of stem cell research is that we could find ways to tell these cells to become any necessary type of cell in order to be used to fix damaged organs, severed spinal cords, etc. In the sense of organ repair, stem cells and cloning techniques may be two different routes to repairing similar damages to the body. Although neither is at present a reliable, effective treatment, one, the other, or both may eventually be used as the situation demands.

Not all stem cells are the same. Pluripotent stem cells are the most useful, because they can differentiate into any of the more than 200 cell types in the body. Unfortunately, the most success in harvesting pluripotent stem cell has come from blastocysts, that is, early stage embryos. This raises a host of ethical concerns over the abortion of early stage fetuses in order to obtain these stem cells, or other means of production that could have possibly led to a viable child if the stem cells were not harvested. Because of these issues, researchers distinguish between embryonic stem cells and adult stem cells, which can be harvested without harm to the individual. A small number of pluripotent adult stem cells can be found in a number of tissues, including umbilical cord blood.

Multipotent stem cells can produce only cells of a closely related family of cells. Unipotent cells can produce only one cell type, but have the property of self-renewal, which distinguishes them from non-stem cells.

Only a few adult stem cell therapies currently exist, such as bone marrow transplants that are used to treat leukemia. However, the potential for stem cells to be used as all-purpose replacement parts to repair any tissue makes the potential for new therapies very promising.

## 27.7 LONGEVITY

There is presently a massive amount of research into medical treatments for almost all of the conditions that are more prevalent late in life. This research is being done owing to the market pressure exerted by the aging baby boomer generation in the United States. Immortality is still in the realm of science fiction novels. However, the average life expectancy is slowly increasing in most countries of the world. It has risen from 28 years in ancient Greece to 67 years in the modern world, and is as high as 82 years in selected regions of the world. More importantly, the quality of life can be kept higher much later in life.

It is unlikely that there will be one wonder drug that extends life. However, research into telomeres, stem cells, genetic manipulation, central nervous system disorders, apoptosis (programmed cell death), and a thousand other things may combine to continue this trend. Environmental factors, such as diet, exercise, and work environment, also play a significant role. In addition to the natural aging process, there are also genetic traits that correlate with having an increased or decreased life expectancy.

Readers are advised to proceed with skepticism when reading up on this topic. Unlike the other topics discussed in this book, the majority of writings on longevity are not works based on validation by the scientific method. On the topic of longevity, there are many publications written as though they were factual, but which in fact range from outright fiction to unsubstantiated theories. Mixed in with this dubious literature are some excellent scientific studies. However, despite the mass of solid scientific work in this area, in a few hundred years history will probably show this work to be early steps of treating the symptoms, rather than the cause.

With social, religious, and moral issues being raised over technologies such as stem cell research, genetic manipulation, and cloning, individuals may indeed have to make a decision whether they are willing to die for those values. Historically, religions have adapted to changing technology, albeit rather slowly. Today, individuals who refuse to undergo surgery or take pills based on their religious convictions are a very small and dwindling group. These new technologies may in time be just as widely accepted and utilized.

## BIBLIOGRAPHY

### Individual Patient Genome Sequencing

Personal Genome Project. Available at http://www.personalgenomes.org/. Accessed 2008 Jan 22.

### Drugs Customized for Ethnic Group or Individual Patient

Helliwell PS, Ibrahim G. Ethnic differences in responses to disease modifying drugs. Rheumatology 2003; 42: 1197–1201.

Kahn J. Race in a bottle: Drugmakers are eager to develop medicines targeted at ethnic groups, but so far they have made poor choices based on unsound science. Sci Am 2007; 297(Aug): 40–45.

Mullin R. Personalized medicine. Chem Eng News 2008; Feb 11: 17–27.

Some Ethnic Groups More Susceptible to Adverse Drug Reactions. ScienceDaily 2006: May 6. Available at http://www.sciencedaily.com/releases/2006/05/060506103939.htm. Accessed 2008 Jan 22.

### Gene Replacement

Pichiorri F, Trapasso F, Palumbo T, Aqeilan RI, Drusco A, Blaser BW, Iliopoulos D, Caligiuri MA, Huebner K, Croce CM. Preclinical assessment of *FHIT* gene replacement therapy in human leukemia using a chimeric adenovirus, Ad5/F35. Clin Cancer Res 2006; 12: 3494–3501.

Roth J, Grammer S. Gene replacement therapy for non-small cell lung cancer: A review. Hematol Oncol Clin North Am 2005; 18: 215–229.

### Cloning

Brown TA. Gene Cloning and DNA Analysis: An Introduction. 5th ed. New York: Blackwell; 2006.

Human Genome Project. Cloning Fact Sheet. Available at http://www.ornl.gov/sci/techresources/Human_Genome/elsi/cloning.shtml. Accessed 2008 Jan 28.

Levine AD. Cloning: A Beginner's Guide. Oxford: Oneworld Publications; 2007.

Most Who Survive Transplants Get Skin Cancer. Science Blog. Fri, 2007-05-18 06:27. Available at http://www.scienceblog.com/cms/most-who-survive-transplants-get-skin-cancer-13254.html. Accessed 2007 May 21.

Wilmut I, Highfield R. After Dolly: The Uses and Misuses of Human Cloning. New York: WW Norton; 2006.

Wilmut I, Schnieke AE, McWhir J, Kind AJ, Campbell KHS. Viable offspring derived from fetal and adult mammalian cells. Nature 1997; 385: 810–813.

## Stem Cells

Bellomo M. The Stem Cell Divide: The Facts, the Fiction, and the Fear Driving the Greatest Scientific, Political and Religious Debate of Our Time. New York: AMACOM/American Management Association; 2006.

Huang Z. Drug Discovery Research: New Frontiers in the Post-Genomic Era. Hoboken, NJ: Wiley-Interscience; 2007.

Lanza R, Thomas ED, Thomson J, Pedersen R, Gearhart J, Hogan B, Melton D, West M, Editors. Essentials of Stem Cell Biology. San Diego: Academic Press; 2005.

Stem Cells. Wikipedia. Available at http://en.wikipedia.org/wiki/Stem_cell. Accessed 2008 Jan 29.

## Longevity

Life Expectancy. Wikipedia. Available at http://en.wikipedia.org/wiki/Life_expectancy. Accessed 2007 May 21.

Longevity. Wikipedia. Available at http://en.wikipedia.org/wiki/Longevity. Accessed 2007 May 21.

Additional references are contained on the accompanying CD.

# APPENDIX

# ABOUT THE CD

This book comes with an accompanying CD. There are a number of things on this CD. One is a file containing the supplemental references. There are also color versions of all of the figures in the book in the subdirectory named "color_figures." A number of the makers of computational drug design software have contributed white papers and product literature, which are also on the CD. The following lists give a simple description of the contents of those directories.

**Accelrys**

- antibody_modeling_app_guide.pdf (white paper)
- catalyst_conformers_0805.pdf (white paper)
- catalyst_proliferation_0205.pdf (white paper)
- de_novo_workflow_app_note.pdf (white paper)
- ds-overview-20.pdf (Discovery Studio product literature)
- electrostatics_task2.pdf (white paper)
- ses-pp-overview.pdf (SciTegic product literature)
- setupdsv201.exe (Discovery Studio Visualizer for Windows)
- setupdsv201.tar.gz (Discovery Studio Visualizer for Linux)
- setupdsvax201.exe (Discovery Studio Visualizer Active X Control)

*Computational Drug Design.* By David C. Young

- student_edition.url (website link)
- wni_ds21_0608.pdf (Discovery Studio product literature)

## ACD

- acd_home.url (web link to ACD home page)
- Downloads.url (link to the free software download page)
- FAQ-Software_Eval_Mistakes.pdf (information on software evaluations)
- logp_vs_logd.pdf (discussion of log *P* vs. log *D*)
- making_sense.pdf (clarification of log *P* prediction results)
- Naming.url (link to the compound naming software information)
- PhysChem.url (link to the physical chemistry software information)

## CCDC_GOLD

- CCDC_gold_suite.pdf (GOLD product literature)
- CCDC_Product_flyer.pdf (list of products from CCDC)
- GOLD_suite_resources.html (links to website)

## CCG_MOE

- HTML (directory with data accessed from the web pages)
- index.htm (product literature)

## COMOlogic

- brochure_Cosmofrag.pdf (product literature)
- brochure_COSMOlogic_overview.pdf (product literature)
- brochure-COSMOmic.pdf (product literature)
- brochure_Cosmosim.pdf (product literature)
- brochure_Cosmotherm_ChemEng.pdf (product literature)
- brochure_Cosmotherm_LifeSc.pdf (product literature)
- brochure_Turbomole.pdf (product literature)
- COSMOlogic_Products.url (product literature web link)
- COSMOtherm_Bibliography_Index.url (references web link)
- demo_download.url (web link to download demo software)

## Conflex

- Barista_Datasheet.pdf (GUI product literature)
- Conflex_Consulting_Datasheet.pdf (consulting information)
- Conflex_Datasheet.pdf (product literature)
- Conflex_Products_Services_Datasheet.pdf (product literature)

## SimBioSys

- eHiTS_on_the_Cell_Whitepaper.pdf (cell processor white paper)
- eHiTS_Whitepaper.pdf (docking white paper)

## Tripos

- BW3DE25-Distribution.exe (Benchware 3D Explorer software)
- Benchware_3D_Explorer_2.3_final.pdf (3D Explorer product literature)

# GLOSSARY

The following are definitions of terms relevant to computational drug design. These definitions are based on common usage in this field. They do not necessarily reflect the dictionary definitions or those in other branches of science.

**3D-QSAR** a method for predicting the activity of a compound when interacting with a specific protein's active site.

***ab initio*** a Latin term for "from the beginning" that describes calculations that are done completely from the quantum mechanical equations, without any parameterization from experimental data.

**active site** the region of the protein or RNA at which it interacts with its substrate or allosteric compound and with drugs.

**active transport** the process whereby natural compounds or drugs are transported to another location by proteins that have that purpose.

**activity** a quantitative measure of how much of a compound is required to have a measured effect on a biological system

**activity cliff** a structure–activity relationship in which a small change in a functional group at a specific location makes a large difference in the activity.

**absorption** the process whereby a drug reaches the bloodstream from the digestive tract, skin, lungs, injection site, or other external locations.

*Computational Drug Design.* By David C. Young

**agonist** a compound the causes a protein to increase its rate of function.

**algorithm** a sequence of steps encoded in a computer program.

**alignment** (of three-dimensional structures) the process of rotating and translating three-dimensional structures such that their shapes are matching on top of one another as much as possible.

**alignment** (of primary sequence) the process of putting two protein or nucleotide sequences next to one another, possibly with gaps, such that identical or similar residues are at the same position as much as possible.

**allosteric** a secondary site on a protein that regulates the behavior of the protein, but does not perform the protein's primary function.

**Ames test** a cell-based assay for mutagenicity.

**anisotropy** refers to atoms within a crystal structure that are in slightly different locations from one molecule to the next within the crystal.

**antagonist** a compound that causes a protein to decrease it's rate of functioning.

**ant colony algorithm** a process that reuses portions of previous steps of the simulation, analogous to ants following the pheromone trails of the ants that traveled that path before them.

**animal trials** tests of drugs in animals to determine the toxicity and efficacy of the drug in an animal.

**Antabuse(disulfiram)** a drug for treating alcoholism.

**antisense drugs** compounds that bind to RNA sequences by having the complementary sequence.

**artificial intelligence** the study of computer algorithms for doing things that would typically require human intelligence.

**assay** an experimental test of a compound interacting with its target (a biochemical assay) or with live cells (a cell culture assay).

**backbone** refers to the carbons and amide bonds that interconnect amino acids in a protein. Also, the core structure of a drug with the functional groups around the periphery removed.

**Bayes' theorem** a statistical analysis that quantifies the probabilities of two events occurring at random, and thus indicates when there is a stronger than random correlation between the events (e.g., certain functional groups and compound activity).

**B-factor** quantifies the positional accuracy of each atom in a crystal structure, due to both thermal motion and the accuracy of the crystallographic analysis.

**bioavailable** the qualitative property of a compound that it will get to the intended target in the body via a particular route (e.g., orally bioavailable compounds can be administered as tablets).

**biochemical assay** an experimental test in which the compound being tested is combined with the target biomolecule, a solvent, and an indicator compound in order to allow spectroscopic measurement of the interaction between the test compound and the target.

**biofilm** a protective layer of carbohydrates excreted by a bacterial colony.

**bioinformatics** refers to a set of algorithms and computer programs for analysis of primary sequences of proteins, DNA, and RNA.

**bioisostere** a functional group that will interact with a biological target in a manner similar to that of another functional group.

**blacklist** a list of compounds or functional groups that researchers want to exclude from consideration.

**blastocysts** early stage embryos.

**blockbuster drug** a pharmaceutical industry term for a drug that has gross sales over a billion dollars.

**blood–brain barrier (BBB)** a layer of epithelial cells that surrounds the central nervous system for the purpose of excluding most blood-borne compounds from entering these sensitive organs.

**Botox** trade name for botulinum toxin type A.

**byte code interpreter** a mechanism by which computer languages such as SVL and Java can be run on multiple computer platforms without being recompiled.

**C++** an object-oriented programming language.

**Caco-2 assay** a test of how well compounds pass through a layer of colon wall cells.

**candidate** a designation indicating how far a drug is through the development process. This designation is typically used late in the development project, such as when a drug is ready to enter animal or clinical trials. The precise criteria necessary to designate a drug as a candidate varies from one pharmaceutical company to the next.

**Carbó Similarity Index (CSI)** a quantitative measure of how similar to molecules are to one another.

**carcinogenicity** the measure of how readily a given compound will cause cells to become malignant tumor cells.

**cardiotoxicity** a measure of how harmful a compound is to the heart, typically owing to hERG protein binding.

**caspase** a protease protein.

**cell-based assay** an experimental measure of drug activity in a cell culture.

**central nervous system** the brain and spinal cord.

**chaperone** a biomolecule that interacts with proteins in order to force them to fold into the correct shape.

**cheminformatics** refers to techniques for storage and analysis of large quantities of data related to chemical structures and properties.

**cherry picking** refers to selecting the best compounds from a list of available compounds to be assayed.

**cisplatin** a cancer chemotherapy drug.

**clinical trials** tests of a new drug in human patients for the purpose of providing enough information to receive approval from regulatory authorities (e.g., in the United States, the FDA) for prescription use.

**clustering** the process of grouping together similar compounds.

**combinatorial library** a collection of compounds all created through a similar synthesis route, and typically on automated synthesis equipment.

**CoMFA** "Comparative Molecular Field Analysis," which is the original 3D-QSAR technique.

**CoMSIA** "Comparative Molecular Shape Indices Analysis," which is a 3D-QSAR technique.

**competitive inhibition** downregulation of a biomolecule target due to a drug compound binding at the site where the native substrate would otherwise bind.

**compound refinement** the process of starting with somewhat active compounds and creating similar compounds with improved activity and pharmacokinetic properties.

**compound selection** the process of using a scientific analysis to choose compounds for assay and compound refinement.

**conformational analysis** the computational creation and testing of multiple conformers of a compound—typically referring to small molecules.

**connectivity index** a single number that describes the linear, branching, and cyclic bonding patterns within a molecule.

**consensus scoring** a process of testing compounds and their poses within a docking simulation based on giving acceptable results as predicted by multiple test functions.

**correlation coefficient** a quantitative measure of how well a value can be predicted as a linear function of a given parameter.

**correlation matrix** a two-dimensional grid of numbers that are correlation coefficients and cross correlation coefficients.

**CPU** "central processing unit," which is the primary chip in a computer, such as a Pentium, Athlon, Opteron or Itanium.

**crevice detection** a computer algorithm for identifying concave regions in the surface of a biomolecules, which might be the active site.

**cross reaction** a side effect that only occurs when two drugs are taken simultaneously.

**crystallography** an experimental analysis of the three-dimensional shape of a molecule.

**cytochrome P450 (CYP)** a class of proteins, mostly found in the liver, that act to functionalize compounds, thus aiding their removal from the bloodstream.

**database** a collection of data in an electronic form that allows it to be searched and sorted.

**decision support software** is designed for helping a researcher analyze a collection of data in order to determine which classes of compounds are the best candidates for continued development.

**dendrogram** a graphical display of relative similarity of molecules or collections of molecules, displayed in a tree-like picture.

***de novo* programs** use artificial intelligence algorithms to automate the structure-based drug design process.

**dielectric constant** the ratio of the magnetic permittivity of a substance to the permittivity of free space.

**distance** (pertaining to molecular similarity) a value ranging from 0 to 1, in where 0 indicates that two molecules are identical, small values less than 0.15 indicate that molecules are similar, and values closer to 1 indicate very different molecules.

**diverse library** a set of compounds in which there is a large variation in chemical structure and physical properties. This term may refer either to data in a computer or to the physical compounds.

**diversity** a measure of whether the compounds in a collection are similar to one another or dissimilar.

**docetaxel** an antitumor drug, similar to paclitaxel

**docking** a computation that simulates the interaction of a compound with a biomolecule's active site, thus predicting the energy and geometry of binding.

**downregulating** decreasing the rate at which a biomolecule performs its function.

**drug** a compound approved by regulatory authorities (e.g., in the United States, the FDA) for use in treating an illness. Also, a compound in the process of design, testing, and approval for this end purpose.

**drug-likeness** a qualitative or quantitative indication of whether a compound could be a drug, usually based on ADMET properties.

**efficacy** the qualitative property of a compound having the desired effect on a biological system. In the case of drug efficacy, this means having a measurable ability to treat a disease's cause or symptoms.

**enrichment** a quantitative measure of whether the number of active compounds assayed from a collection of *N* compounds selected by a given software tool (e.g., docking or pharmacophore search) is better than the number of active compounds that would have been identified if *N* compounds had been selected at random.

**enumerative algorithm** refers to a software package designed such that every compound is generated individually in software to allow analysis and prediction of properties.

**enterprise information systems** a central data repository that makes it possible for any authorized employee of a large organization to access experimental and computational results.

**epothilone** a macrocyclic natural product that is active as an antitumor drug.

**excretion** the process by which unwanted compounds (e.g., waste or potential toxins) are removed from the bloodstream.

**expert system** an artificial intelligence algorithm based on codifying the thought processes of people who are experts in the applicable field.

**Fen-Phen** a diet drug that was removed from the market because of severe cardiotoxicity side effects.

**fingerprint** an array of numbers that describes a molecule in a way that can act as a basis for comparing molecular similarity.

**focused library** a collection of similar compounds sharing a structural motif and exhibiting variations in side chains and functional groups.

**force field** a set of parameters and equations that use those parameters, which is a specific implementation of a molecular mechanics method.

**Fortran** a programming language designed for mathematical and scientific applications.

**frequent hitters** *see* promiscuous binders

**fuzzy logic** an artificial intelligence algorithm that encodes "maybe" results as percentages or probabilities.

**Gaucher's disease** an illness caused by an enzyme deficiency.

**genetic algorithm** a way of finding a near-optimal solution to a multivariable problem by constructing a simulation with a population of solutions that can have children, mutations, etc. analogous to the natural selection process.

**Guénoche algorithm** a specific hierarchical clustering procedure.

**genome** the collection of DNA sequences in an organism. Also, the sequences of the proteins generated by translation of the open reading frames. Often refers to this data stored in computer-readable form.

**geometry optimization** the processes of changing the shape of a molecule to find the most energetically favorable three-dimensional shape.

**GLUT2** the glucose transporter protein.

**GPCR** "G-protein-coupled receptor," which is a target for many drugs.

**grid potential** (docking) a method for making docking calculations run faster by pre-computing a portion of the mathematical analysis on a regularly spaced collection of points in the active site.

**grid search** (conformational analysis) an algorithm in which each rotatable bond of a molecule is tested at each of $N$ positions, and all combinations of conformations of each bond are tested.

**group additivity** an algorithm for predicting molecular properties based on numerical weights assigned to each functional group.

**group average** a nonhierarchical clustering algorithm.

**half-life** the amount of time before the concentration of a compound in the bloodstream decreases to half of its peak value.

**heuristic algorithm** an artificial intelligence process in which the route to obtaining results is guided by simple rules of thumb, called heuristics.

**hierarchical clustering** an algorithm that gives multiple ways of dividing a group of compounds into clusters, related by a parent–child relationship.

**hepatotoxicity** the quantitative measure of how much a drug will damage the liver.

**hit** refers to a compound that was found to have some activity towards the desired target, typically at early stages of the screening process.

**homology** the amount of similarity between two protein or DNA sequences. Also, a hypothesis that the two proteins share an evolutionary ancestor.

**homology model** a three-dimensional model of a protein constructed by comparing the primary sequence with that of proteins for which crystallographic structures are available.

**HookSpace method** an efficient algorithm for analysis of the shape of possible conformers of large collections of molecules.

**hPEPT1** "human intestinal small peptide carrier," which is responsible for the active transport of a significant number of drugs.

**hydrophilic** refers to the qualitative or quantitative description of how well a compound will dissolve in water.

**icon-oriented programming** a process in which a computerized procedure can be defined by selecting graphically displayed boxes and connecting them by dragging a mouse to indicate the flow of data from one to the next.

**identity** the extent to which two primary sequences are identical, typically expressed as a percentage.

**InChI (IUPAC International Chemical Identifier)** a way of expressing chemical structures unambiguously, thus making it easy to determine computationally if two compounds are identical, enantiomers, tautomers, etc.

**induced fit** a process by which the shape of a protein's active site changes depending upon the ligand bound in that site.

**information content** a mathematical measure of molecular complexity.

**inhibition** a reduction in the rate at which a protein performs its function.

***in silico*** Latin for "in silicon," meaning that it is a value generated by a computer program (computer chips are made from silicon).

**intellectual property** a legal ownership of the rights to use and sell an idea or product, assigned based on a patent, copyright, or trademark.

**intrinsic reaction coordinate (IRC)** the lowest energy route from a transition structure to the reactants and products in a chemical reaction.

***in situ*** Latin for "in the original position," implying an experiment done in a living animal, or sometimes in live cells.

***in vacuo*** Latin for "in vacuum," indicating that a calculation or experiment is done without the presence of a solvent.

***in vitro*** Latin for "in glass," indicating a physical experiment done in a laboratory (in glassware, plate, etc.) Biochemical assays are considered *in vitro* assays. Cell-based assays are sometimes considered *in vitro* and sometimes considered *in situ*.

**Jarvis–Patrick** a simple algorithm for nonhierarchical clustering.

**Java** a programming language made to run graphical interfaces on multiple platforms without recompiling.

**Jython** a Java implementation of the Python programming language.

**Kier and Hall indices** single numbers that describe the shape of a molecule.

***K*-means** a nonhierarchical clustering algorithm.

**knockout animals** living animals that have been genetically manipulated to remove one open reading frame from their genome.

**knowledge-based algorithm** *see* heuristic algorithm

**Kohonen map** a two-dimensional graphical representation of a multidimensional space of data.

**lead** a molecular structure motif that has been selected as the basis for more intensive drug design efforts.

**lead hopping** refers to a process that can find active molecular structure motifs that have not yet been identified for a given target.

**lead-likeness** sometimes synonymous with drug-likeness; sometimes based on more restrictive criteria than drug-likeness; sometimes based on physical properties of molecules, as opposed to drug-likeness being based on structural features.

**library** a list of compounds. May refer either to physical samples of the compounds, or a list of data in a computer.

**ligand-based drug design** refers to drug design processes that can be used when the biomolecular target is unknown.

**linear regression** a mathematical process for getting the best possible fit of parameters to predict some property.

**Linux** a widely used computer operating system. A public domain clone of the Unix operating system.

**Lipinski rule of fives** a set of criteria for predicting if a compound will be orally bioavailable.

**lipophilic** refers to the qualitative or quantitative property that a compound can be dissolved in lipids.

**lock-and-key** refers to a theory of drug design that the drug should fit in the target's active site like a key fitting in a lock.

**Manhattan distance** a worst case measure of chemical similarity.

**matrix least squares** an efficient method for finding coefficients that allow parameters to be used to predict a property. Mathematically equivalent to linear regression.

**maximal common subgraph (MCS)** refers to a pattern of atoms and bonds that an entire collection of compounds all contain.

**mechanism** refers to the detailed description of a chemical reaction. This includes geometric motion of atoms, energetics, and sometimes vibrational motion.

**memory** refers to the amount of data that a computer can hold in silicon chips, without accessing the disk drive.

**metabolic blocking** refers to adding functional groups for the purpose of preventing a compound from being metabolized by a given mechanism in the body.

**metabolic pathway** the sequence of proteins and small molecules upon which they act that perform some function within the cell.

**metabolization** the process by which proteins alter compounds in order to use them or to remove them from the body.

**metalloenzyme** a protein with a metal atom in its active site.

**metascale** a modeling technique for working with systems (e.g., cells, membranes, or organs) that are too large to be modeled as a collection of individual atoms.

**micromolar** refers to a compound with a weak activity, of the order of $10^{-6}$.

**MOGA (Multiobjective Genetic Algorithm**) a specific process for performing multiobjective optimization.

**molecular dynamics** a simulation that shows how molecules move, vibrate, diffuse, and sometime react over time.

**molecular electrostatic potential (MEP)** the mathematical computation of regions around a molecule at which positive or negatively charged species will bind.

**molecular mechanics** a method for simulating molecules with the atoms represented as weights connected by springs and having a partial charge.

**Monte Carlo algorithms** test many possibilities based on a random number.

**multidimensional optimization** *see* multiobjective optimization

**multiobjective optimization** refers to a process for finding a solution that is optimal in terms of multiple criteria.

**multipole** a representation of the charge distribution in a molecule, such as a dipole, quadrupole, octupole, etc.

**mutagenicity** the qualitative and quantitative tendency for a compound to cause mutations in the genome of cells, particularly during mitosis.

**nanomolar** refers to a compound with a potent activity, of the order of $10^{-9}$.

**native substrate** the species that would be acted upon by a protein in the course of the normal functioning of the body.

**neural network** an artificial intelligence algorithm that predicts something based on a simulations of the way in which the neurons interact in the brain.

**neurotoxins** compounds that are toxic owing to their ability to interrupt nerve impulses.

**non-enumerative algorithm** refers to a process that analyzes entire collections of molecules without ever creating a structure for each individual compound.

**nonlinear map** a way of displaying molecular similarity by representing similar molecules as points near one another on a map, and dissimilar compounds as points far apart.

**NP-complete algorithm** a class of mathematical problems that require a very complex, time-consuming computer algorithm to solve.

**oral bioavailability** the property of a drug that it can be swallowed in solid or liquid form and will be able to reach its intended target in the body.

**paclitaxel** an antitumor drug, similar to docetaxel

**Pareto optimization** a multiobjective optimization algorithm analogous to a downhill energy search.

**partial least squares** a mathematical process for finding a way of predicting activity based on some collection of parameters.

**passive intestinal absorption** the process of a drug passing from the intestine into the bloodstream because it is sufficiently lipophilic to diffuse through the intestinal wall.

**patent** a legal document granting a person or company ownership and exclusive rights to sell something for a number of years.

**pathogen** a harmful virus or bacteria.

**peptidomimetic** a compound that is similar in shape to a peptide.

**pharmacokinetic** the properties of a drug in terms of its absorption, distribution, metabolism, and excretion.

**pharmacophore** an arrangement of interaction features (e.g., hydrogen bond donor/acceptor, steric, etc.) that can be used to search for compounds that might be active.

**picomolar** refers to a highly active compound, with activity of the order of $10^{-12}$.

**primary sequence** the order of amino acids in a protein chain, or nucleotides in a DNA or RNA chain.

**principal components analysis** weights of parameters that define a basis for prediction of activity.

**privileged structures** *see* promiscuous binders

**polarity indices** numbers that describe the charge distribution in a molecule.

**polarizability** the readiness with which electron charge density shifts from one part of a molecule to another.

**prodrug** a compound that releases an active drug once it has been metabolized in the body.

**promiscuous binders** compounds that bind to many proteins or give false signals, thus causing them to test positive in assays for many different targets.

**protein folding** the process of attempting to compute the three-dimensional structure of a protein based on the primary sequence alone.

**proteome** the collection of all of the proteins in the body.

**proteomics** the study of the proteome.

**Python** an interpretive programming language.

**QSAR** "quantitative structure–activity relationship," a way of predicting molecular properties.

**quantum mechanics** mathematical law of nature that describes the behavior of electrons, and thus chemical bonding.

**Quantum Similarity Measure (QSM)** a metric for comparing molecules.

**Randic indices** numbers that describe the shape of a molecule.

**rational drug design** *see* structure-based drug design

**reaction coordinate** geometric and energetic description of a chemical reaction.

**receptor** a protein at the surface of a cell that passes a signal into the cell when an appropriate molecule binds to it. Also, the active site of a protein, where a signaling molecule binds.

**resistance** change in a pathogen that makes it less susceptible to a drug.

**resolution** the accuracy of a crystallographic structure.

***R*-factor** a numerical measure of the resolution of a crystallographic structure.

**rotamers** another name for conformers.

**SAR** "structure–activity relationship," a qualitative description of some constraint on the molecular structure needed for activity.

**scaffold hopping** the finding of previously unknown structural motifs that are active against a given target.

**Scientific Vector Language (SVL)** a chemically aware programming language.

**scoring** (in docking) a numerical measure of how well a molecule binds to a target's active site.

**scoring plates** the automated spectroscopic analysis of experimental assays.

**screening** the experimental measurement of compound activity. Synonymous with assaying.

**script language** an interpretive computer programming language.

**semiempirical** quantum mechanical methods that substitute some of the most time-consuming integrals with parameterized values.

**serotype** a mutated form of a pathogen.

**side effects** unwanted physiological symptoms from taking a medication.

**Silvadene (silver sulfadiazine)** an antimicrobial drug.

**similarity** numerical measure of how closely two molecules compare with one another.

**simulated annealing** an algorithm for finding a near-optimal solution to a global optimization problem.

**singleton** a cluster with only one compound in it.

**SLN** "SYBYL Line Notation," a full featured string format for representing chemical structure.

**SMILES** "Simplified Molecular Input Line Entry Specification," a string format for representing two-dimensional chemical structure.

**specificity** the property of a drug that it inhibits the intended target, and not any other proteins in the body.

**statistical mechanics** the discipline that attempts to relate the properties of macroscopic systems to their atomic and molecular constituents.

**strategic bonds** chemical bonds that are targeted as the points to attach or remove functional groups as part of a synthesis route.

**structurally conserved regions** pieces of proteins, for which crystal structures are available, used to construct a homology model.

**structural motif** the central part of a drug's chemical structure. Many different functional groups will be added to this core region to generate focused libraries.

**structural unit analysis** an algorithm for generating structure–activity relationships.

**structure-based drug design** the process of designing drugs for a target for which the three-dimensional structure is known.

**substrate** a small molecule or small section of a large biomolecule that binds in a protein's active site.

**substrate analog** a drug that is designed to look similar to a target's native substrate.

**substructure** a piece of a chemical structure, typically used to search for other molecules containing the same pattern of atoms and bonds.

**suicide inhibitor** a drug that binds irreversibly to its target.

**support vector machine** a learning algorithm for creating a system to predict some property.

**synthetic accessibility** a measure of how easily a compound can be synthesized.

**synthons** a functional group or molecular fragment used to generate a combinatorial library.

**tabu algorithm** a process that searches a space (e.g., molecular positions in a docking calculation), but excludes sections of the space that have already been searched.

**Tanimoto distance** the most widely used measure of molecular similarity.

**target** the biomolecule in the body with which a drug is intended to interact.

**targeted library** *see* focused library.

**template** (in homology model building) a similar protein with a known three-dimensional structure, used to guide the creation of a homology model for the unknown protein.

**teratogenicity** a form of toxicity resulting in a malformed fetus if taken during pregnancy.

**thalidomide** a failed drug, which caused severe birth defects.

**threading algorithm** a process for aligning marginally similar protein structures for the purpose of creating a homology model.

**time complexity** the way in which the required CPU time for a calculation scales with the size of the problem.

**tubulin** a protein that forms microtubules during cell mitosis.

**topical** application of drug to the skin, for example in the form of a cream.

**topological** mathematical description of shape, such as connectivity, branching, and rings.

**topological polar surface area (TPSA)** a molecular descriptor frequently used to describe how a molecule will interact with a polar or nonpolar environment.

**toxicity** the quantitative quality of a compound doing harm to the body.

**transgenic animals** have an extra gene from another species inserted in their genome.

**transition state/structure** the three-dimensional structure and electronic structure of a molecule at the transition point (energy maximum) in a chemical reaction.

**transition state analog** a drug designed to look like the intermediate of the reaction of the enzyme that it is inhibiting.

**uncompetitive inhibition** occurs only when an inhibitor binds to the enzyme–substrate complex.

**Unix** a computer operating system used for workstations, servers, and supercomputers.

**upregulating** the process of a drug increasing the rate at which the target performs its action.

**variable regions** sections of a homology model for which no template is used.

**Viagra (sildenafil)** a drug for erectile dysfunction.

**Ward's algorithm** a hierarchical clustering process.

**Weiner index** a number that describes the shape of a molecule.

# INDEX




*Computational Drug Design*. By David C. Young